Morbid

Morbid

Debunking Modern Longevity Science

Saul Justin Newman

The MIT Press
Cambridge, Massachusetts
London, England

The MIT Press
Massachusetts Institute of Technology
77 Massachusetts Avenue
Cambridge, MA 02139
mitpress.mit.edu

The MIT Press would like to thank the anonymous peer reviewers who provided comments on drafts of this book. The generous work of academic experts is essential for establishing the authority and quality of our publications. We acknowledge with gratitude the contributions of these otherwise uncredited readers.

This book was set in ITC Stone Serif Std and ITC Stone Sans Std by New Best-set Typesetters Ltd. Printed and bound in the United States of America.

Library of Congress Cataloging-in-Publication Data

Names: Newman, Saul Justin, author
Title: Morbid : debunking modern longevity science / Saul Justin Newman.
Description: Cambridge, Massachusetts : The MIT Press, [2026] | Includes bibliographical references and index.
Identifiers: LCCN 2025041946 (print) | LCCN 2025041947 (ebook) | ISBN 9780262052719 hardcover | ISBN 9780262052726 pdf | ISBN 9780262052733 epub
Subjects: LCSH: Longevity | Aging--Prevention | Fraud in science | Blue zones
Classification: LCC RA776.75 .N49 2026 (print) | LCC RA776.75 (ebook)
LC record available at https://lccn.loc.gov/2025041946
LC ebook record available at https://lccn.loc.gov/2025041947

10 9 8 7 6 5 4 3 2 1

EU Authorized Representative: Easy Access System Europe, Mustamäe tee 50, 10621 Tallinn, Estonia | Email: gpsr.requests@easproject.com

To Arthur and Leila, Tiger and Squid, and the rest of my wild family—
Ryan, Hannah, Ellen, Annika, Keith, Maree, and all the wriggly brats.

Contents

Preface: Grave-Robbing in the South of France

It was 10:00 a.m. on a Tuesday when the email to Vladimir Putin came through. The Queen was cc'd too, presumably in consideration of her balanced and worldly opinions. Following her majesty was a shockingly large collection of 148 famous scientists, demographers, and a bizarre mix of grandees. At the top of the list, as the recipient, was my name.

Dr. Zak wanted to dig up the corpse of the world's oldest woman, Madame Jeanne Calment, and he was asking me to help.

Dr. Zak was of the determined opinion that the world's (supposedly) oldest woman had to be exhumed despite having lain in the rich French soil for the best part of two-and-a-half decades. Some vials of her blood had gone missing, and it was apparently imperative to dig up her corpse and check if it was inbred. Dr. Zak had naturally thought it necessary to enlist the help of Her Royal Majesty, me, and a bemused mass from across academia. Without a whiff of context, subtlety, or preamble, he had drafted a plan for us to do some grave-robbing and hurled it like a burning phosphor bomb into the emails of some very busy people.

It was a rather interesting change of course from my average Tuesday.

It was also, I thought, not the worst idea. Paperwork aside, it would only take a few thousand dollars, personal protective equipment, sterile cotton swabs, and a shovel. What was the worst idea, of course, was the horrifying email chain of professorial superstars and literal royalty who were on the receiving end of this bizarre missive, with my name front and center as an apparent coconspirator.

Having only just sat down with my morning flat white, the collective replies of eight time zones and a hundred or more angry professors promptly kicked off.

The first to really get started were the accidental reply-alls. Emeritus professors, the semiretired benevolent grandparents of the academic world, naturally began a near-exponential cascade of accidentally pressing "reply-all" to send their apologies and say that they did not have the faintest notion what we were talking about. These polite requests to be left alone were, again very naturally, followed by several of the "full superstar" professors (or their assistants) angrily demanding that everyone *definitely not* send any further reply-all messages to the thread. Given their wish to inform the entire email chain, so as to avoid further messages from anyone, they naturally spread this nugget of wisdom using a further avalanche of reply-all messages. By this time, our collective emails were probably consuming the annual electricity output of Botswana.

After this, the inevitable public hounding began. For the uninformed, the supercentenarian database is a glorified collection of spreadsheets containing second-hand records of newspaper reports on the world's oldest people—people who survived to age 110 or over—collectively known as "supercentenarians." The head of the supercentenarian database was wholeheartedly devoted to the venerable and, he thought, near-holy duty of recording the ages and personal foibles of these grandparents of humanity, and he was not precisely happy. His response to the possible disinterring of the oldest woman in the world by a Moscow mathematics graduate was less than ideal, containing whole paragraphs of vehement angry responses conveyed in nonstop capital letters. The case was INDISPUTABLE. The records were VERIFIED. The criticisms were FANTASY. Much was then said about me, at extraordinary length and speed, especially how I too had besmirched the dignity of the oldest people in the world and was virtually a criminal. Quite true, I thought to myself, but hardly very relevant.

Having communicated with this gentleman before, I knew it was his usual style to write with such naked anger that the reader thought the writer's impending coronary rupture would be visible from Jupiter's moons. While I had acclimated to this writing system, the spectating academics (and, I incorrectly imagined to myself, the amused Queen) were making their first engagement with this challenging communication style. This promptly caused the reply-all errors to momentarily propagate themselves at an even higher pace, which in turn accelerated the already blistering pace of bizarre accusations of fraud and lying from the head of the supercentenarian database. Unfortunately, as I had been personally named several times in colorful language, the thread now required some response from me.

First, however, I had to consider recent events.

The would-be junior partner in this Russian Burke-and-Hare caper, in which we would excavate a rather far-gone French corpse, was Dr. Nikolay Zak. Dr. Zak was a mathematician turned glassblower living on the edge of Moscow, who had recently been interviewed by *The Washington Post* after explosively claiming that the identity of the world's oldest woman was false. Dr. Zak had presented a pile of evidence, to which I had hardly paid attention, but had supplied some mild and wholly private support. This evidence pointed Dr. Zak to a quite simple and surprisingly common occurrence to explain the longevity of Jeanne Calment: that somehow, Jeanne Calment's daughter had adopted her identity.

The whodunit certainly had appeal. Jeanne Calment was rich, French inheritance taxes were higher than the Pyrenees, and her daughter was one of the long-term unemployed. Added to these simple incentives was an unusual extra: Jeanne Calment had preemptively sold her home, which was also her daughter's abode, under an unusual French contract. In this contract, the buyer pays a hefty allowance, usually to an elderly individual, in exchange for the legal title of their home when they die. As such, had Madame Calment's daughter been alive at her mother's passing, the law would make her homeless, give at least a third of her mother's wealth to the state, and split the remainder of her inheritance in half with her long-separated brother. Good motives for a little light-fingered fraud, indeed. However, such details are spoilers: you will find out about the world's oldest lady later, after I wade through the emails.

Dr. Zak had now happened upon some half-forgotten rumor or record of a vial of Madame Calment's blood, drawn during some late stage of her life and subsequently held in the drawers of some would-be lifespan extension company. Clearly working himself up into a state of grandiose scheming, he had correctly concluded that genotyping this vial of her blood would reveal her inbred status. He had, unfortunately, also concluded that digging up and testing Calment's moldering husk was a viable second option—one he was now pursuing.

The outcome of this was that, unlike the bemused social media managers at the Kremlin and the Royal Palace, I needed to express an opinion in that email thread. Unfortunately, my opinion was that digging up Jeanne

Calment was actually fairly rational. Exhumation and genotyping could solve the problem of her identity. The challenge, of course, was to convey this in a manner that painted me in a professional light rather than as a marginally insane grave robber. I reverted to flimsy humor, ignored the madness of the supercentenarian database manager, and offered to buy the genotyping kit for the corpse, if I could only be removed from the email chain. I was also at pains to point out that I was sorry that such an extraordinary number of people had been inconvenienced, and attempted to convey that I was not the magician in the show but the lady being sawn in half.

It hardly worked. The furious email exchanges accelerated.

However, among the electronic shouting, I had spied a quantum-dot of hope. Wedged between the theories of young Dr. Zak and the mad-as-an-electrocuted-badger rantings of the incendiary Robert Young—the man who ran the supercentenarian database—an odd tidbit of information had glued itself to me. I had heard the Calment stories before. I had heard all the strange rantings already. Yet, there was one thing entirely new to me. What was this vial of blood?

The question led me on a short chase through Google. At some time around 2002, it seemed, one Dr. François Schächter had founded a company to solve the problem of aging. In my line of work this is a routine event, a steady habit since the sixteenth century, in which a confident mountebank transitions his company from a brash start-up to a quiet and inevitable failure. As we'll see later, the best mountebanks have an overweight mattress stuffed with other peoples' money by the time they reach the failure stage. François, however, appeared to have noble intentions and had gone through the "quiet failure" process with his integrity wholly intact. He had also, he claims, collected Jeanne Calment's blood in the process.

After further digging, I found that François's company had long gone under or evaporated, with a long-dormant website last scraped in 2017, and the vials of blood now rested with an institute called the Fondation Jean Dausset in Paris. The rumored vial of blood that Dr. Zak had mentioned was real and, it seemed, there were two vials: more than enough for some genotyping work. No corpse excavation was required. Why be a graverobber when you can be a vampire?

This little fact, thank goodness, gave me an exit strategy from an email chain that was now so large you could have it printed and bound. I fired off a joke, with a price list of mail-order genotyping kits ($110 for your family or friends), and quit the stage entirely. I wandered outside, picked up an Italian-style chocolate-filled donut and another flat white, and pondered what Putin had made of all this. I did not even know what to make of it myself.

While the coffee percolated my interior, I thought about the case. I am a gradual thinker, like the hobbit Farmer Cotton in *The Lord of the Rings*: slow on the outside but always attempting to think my way through a brick wall on the inside. As I pondered and digested, I realized the short Google chase turned up results, and the results were annoying me.

To know why requires a two-part microlesson. Part one is the family of Jeanne Calment: Calment married her double-second-cousin (on both sides), which leaves a genetic trace in your children and makes it possible to identify if the blood samples were taken from the daughter. The second half of the microlesson rests on a new statistical process. We know the family tree of Jeanne Calment, and we know she has distant surviving relatives: they could agree to give us blood samples, if we ask very nicely. Through a novel technique called a "long-range familial search," we could then therefore discover if the blood in the vials matches her relatives.[1] In short, anyone with Jeanne Calment's blood can check if that blood belongs to an unrelated impostor.

The Fondation Jean Dausset is named after a Nobel Prize–winning immunologist who made a genetic discovery. They own a genotyping lab. And they have the blood of one of the most famous people in the world, who is famous *for a biological reason*. The institute, as is clear from their later work on her pooled genotype, even has ethical clearance to genotype Jeanne.[2] Why would they refuse to confirm they even had the sample?[3]

In science, unscoopable and newsworthy results are an incredible treasure. They can be career-making. They can fund an entire institute. So why, I wondered, would anyone sit on something as invaluable as the genome of the world's oldest person, for decades, *especially* when it is possible to independently validate if the sample was real? Why had the Fondation Jean Dausset not published the genome of the world's oldest woman in a top journal like *Science* or *Nature*, scooped up the grant money, and popped open the Chablis? It was easier and cheaper than getting a corpse-digging permit, and far more rewarding.

These ideas were building a bigger picture for me, adding to the process of steadily thinking my way through the brick wall. I was utterly indifferent to the question of whether the world's oldest woman was a fake. If she was the daughter who escaped paying inheritance tax and became world famous in the process, bonne chance to her. If she was a strikingly weird outlier from a biological distribution, again, happy days, big deal. What disturbed me was a much broader problem: the lack of anything like a functioning research culture in what I had just seen.

It was obvious that few in the mass of unintentional bystanders, busy professors, Nobel laureates, and pooh-bah dignitaries on that email chain cared about Jeanne Calment or the questions raised in the thread. However, for the devotees of aging science in that email chain, it disturbed me that only one had asked, and none had answered, such a basic question as "why don't you just genotype the blood sample."[4] Instead, a reasonable question by Dr. Zak, a likably odd and unfortunate amateur in the field, was met with psychotic hostility.

The way these people reacted to simple questions was, well, madness. One gerontologist reached such a state of tantrum that they resigned because an outsider asking a question "disgraced the field of gerontology."[5] Most disturbing of all was the collective reaction to testable solutions that would benefit everyone. When I had dared suggest an inexpensive opportunity to genotype Jeanne Calment, and scoop up a great research paper, the assembled scientists displayed a research culture so unhinged I felt like I was watching someone eat a baby flambé.

The question was, why?

Unfortunately, the hostility of aging researchers was going to be overtaking my life in a way that made the answer very clear. I was to be publicly accused of spying, fraud, lying, stupidity, and hidden agendas. In private, my intellectual adversaries would attempt to get me fired as a plant scientist. I would then be made a research fellow at Oxford University, only to be isolated, shouted at, and (worst of all in British society) politely ignored. Eventually, despite it all, I would discover a remarkable scandal.

Yet for now, I was going to remain stuck to demographers against my will—like a mouse on a glue trap—because I had just discovered that tens of thousands of the world's oldest living people were secretly dead.

1 Zero Is One

In late 2016 a friend had sent me a scientific paper that he thought I would find interesting.[1] I hadn't talked about aging in a while, but somewhere along the line my mate, and fellow victim of a medical science PhD, had correctly understood my interest in the subject. So, I found myself diverted.

The paper was an absolute shocker, and it was published in *Nature*.

For those fortunate enough to be uninformed, a scientist publishing a discovery in *Nature* is like a journalist publishing a major scoop in *The New York Times*. It is a big, career-making deal. Old friends will email you, and bitter enemies will clench their teeth and squeeze out a "well done." So, it was with some disbelief that I realized this *Nature* paper was utterly defective.

For a while I buried the thought. I was one year into my first postdoctoral job: yet another make-or-break step in a scientific career, when you have no papers, no energy for great post-thesis efforts, and the mortifying feeling that your contract is too short. I was cracking on with plant science, as best I could, in the Black Mountain laboratories of the Australian government. I had a knack for statistics and cross-discipline research, and I had been busy creating something crushingly difficult: by linking genomic data on plants with farm records, theoretical models, and satellite images, I was creating a machine learning model of plant yield and survival.[2] In short, I was training an artificial intelligence (AI) to predict crop behavior and help us feed people. The idea was not just to make a prediction, however, but to take this statistically demanding task and go even further: by writing thousands of lines of code, I was working at understanding how the AI models were making decisions.[3]

Compared to this day job—which required knowledge of statistics, genomics, computation, plant behavior, and satellite imagery—demography was a pleasant distraction, something I had picked up as an interest during my medical science training and probably should have sidelined. But the paper nagged at me. It was so hauntingly bad that it demanded attention, like a rich-country tourist in a poor-country beach resort.

After a few weeks I crumbled. Out of combined procrastination and idleness, I began to make a list of errors, and errors, and errors. By the time the list of stuff-ups exceeded the length of the original paper, I knew I had to do something.

The topic of the paper was interesting, and the conclusions reached were absolute. Dr. Xiao Dong, Dr. Brandon Milholland, and Professor Jan Vijg had penned their thoughts on the upper limits of human life—namely that humanity had reached an upper limit of survival, an age beyond which we could not survive—and they had done so using two main observations.[4]

First, these authors took all the mortality data they could find and used these data to claim that survival was improving across history at all ages, except among the very, very old. Life expectancy had not improved for the oldest old people because, they reasoned, the oldest old were pushing up against the limits of survival.[5]

Second, they had observed the changing records of the oldest people in the world. Their analysis showed that the record for the oldest living person increased, peaked with the oldest woman in the world (Jeanne Calment), and then flattened out or even retreated. It was also, they understood, a sign that humanity was bumping up against an upper limit of survival. We could live to a limit of about 115–125 years and no more.

The only problem was that both arguments were based on mistakes.

Human life is often measured by something called a "life table": a kind of spreadsheet that measures death rates, used by your insurance company and other weirdos. In the life table are two closely related columns. One records the odds of dying at a given age: for example, if you have a 1 percent chance of death at age fifty, the value in row 50 will read 0.01. Another column charts the number of people who would, out of a hypothetical pool of 100,000 babies, survive to reach that age. When the odds of survival to old age fall below 1 in 100,000, this second column is rounded down to zero: obviously, because your odds of becoming that old are less than 1 in 100,000. But your odds of dying at that age will not be zero—the actual odds of survival will be embedded in the *other* column, the first column.

The authors had analyzed data from the wrong column, the one that gets rounded off to zero.[6] Not content with this achievement, they then turned all these zeros into ones. It seems they managed this by charting zero on a logarithmic (log) scale, by assuming that log-transforming zero will give you zero on a log scale. For the math challenged, the log-transform of zero equals negative infinity, and zero on a log scale is equal to one. So, they accidentally made their zeros equal one.

These authors seem to have landed in the most whizz-bang best-in-the-world scientific journal because they rounded off almost all their data to zero, accidentally made these zeros into ones, and concluded that data almost entirely composed of a long string of ones doesn't change much over time. Dazzling stuff.

Over several months I reconstructed this analysis, line by painstaking line of code, and reproduced the glaring blunder.[7] If I ran the analysis using the data rounded off to zero, I could recreate every one of their two-hundred-plus figures and get the exact same results as the original discovery. If I reran the analysis, using the correct column from the exact same spreadsheet, I naturally got the opposite result.[8] Life expectancy at age 110 was improving faster than at any other age, not slower, and there was no sign of an upper limit to human life.

The second major error was even more tragic than the first. The authors' second claim was that the record for the oldest human alive had hit a wall and was no longer increasing: in fact, it had risen to a peak around 1995 and was now going down over time.[9] A great conclusion, unless you considered that the size of the population being measured was also going up and then down over time.[10] The keepers of longevity records had added new countries to their database until 1995. Then, from 2003 onward, they stopped covering countries in the survey, reducing the observed population size over time.

That was it. That was the answer. The maximum record for longevity had not hit an upper limit and stopped rising: the people keeping the longevity records had started and then stopped including major countries in the database. It was like "discovering" that your boat slowly rose and sank at a dock, without suspecting the tide.

It was also, among 2.7 million competitors, the eleventh most-discussed scientific paper of 2016.[11]

I discovered a swarm of other problems in my slow reanalysis, and these doubts began wearing away at the resolve of my PhD supervisor and now coauthor Professor Simon Easteal, who I had turned to for advice when I first read the paper. To carefully recreate the original supplementary figure 6, for example, I had to download the wrong Excel spreadsheet, commit a basic coding error that deleted every death from May and June, and plot the data with an incorrect axis.[12] Eventually, I convinced Simon and we submitted the paper to *Nature*, after which we received a curt reply to the effect that "we have had enough criticism, we don't want any more, please send your paper elsewhere and goodbye."*

Through delays and doubts, but also because of my slow and careful recreation of every single figure and finding, I had missed out on my own *Nature* paper. Five other labs had beaten me to it.[13] None of these labs spotted the cataclysmic problems I had seen: they had instead grabbed one or two other problems with the paper, from a substantial pile, and rushed their findings to press. I berated myself for not doing the same. Speed is more important than being careful in the reward system of modern science.

However, while I mourned my missed opportunity over lattes and self-pity, a public bun-fight had begun.

Reviewers are essentially a tiny science jury: one that often knows you personally, competes with you for grant money, and decides if your scientific paper gets published, executed, or sent back to review-jail. Like jurors in a court, reviewers are also supposed to be fair and independent, and the editor is supposed to act like a judge.† It is an honor system, largely performed for free, that struggles to function.

After five labs slammed the "limits of human life" paper, some questions were asked about the review process, and everyone became a bit saucy. Early on, an award-winning news article had immediately raised the stakes to "deathmatch" when it exposed the degree of bias and bad practice[14] that seemed to have pushed this paper across the line.‡ The paper was initially rejected by the editor, had come back after substantial involvement by the reviewers, and was rejected again. After the paper was rejected with an unequivocal "no" the second time, one of the reviewers had teamed up with one of the study authors and asked the editor pointedly "why don't you take another look?"[15]

---

*This was couched in some polite, arcane, and euphemistic language, but the message seemed clear enough.

†If those two descriptions sound contradictory and open to abuse, well, yes.

‡Though this has naturally had no effect on the subsequent career of the authors or the paper's continued citation.

This supposedly independent reviewer, Professor Olshansky, had been pushing the idea of a lifespan limit for decades. He even claimed that "I knew the data better than the authors. . . . I knew the conclusion" without reading the paper.[16] Of the other two juror reviewers, one was ambivalent but seemingly half-asleep, confessing "I don't remember whether I looked at [the appendix]" because it was "boring, noisy, and confusing."[17] He had initially rejected the paper, before changing his mind after the authors agreed to use his database. The third reviewer was vehemently against the paper being published but was bluntly overruled. The jury system had failed, and the paper was pushed through.

The postpublication assessments were less sleepy. A prominent demographer—Professor Jim Vaupel—summarized the mood when he called the paper a "travesty in a century-long saga"[18] where "they just shovelled the data into a computer like food into a cow."[19] He backed this up with the opinion that "the 'findings' stem from their naive use of inappropriate data . . . they contribute nothing useful."[20]

When told that such criticisms were bound for publication, the study's lead author Jan Vijg revealed his thinking to a reporter: "'I hope *Nature* accepts one of them,' says Vijg on the phone. Because then he'll get to write a response. 'And we'll have another article in *Nature*.'"[21] Vijg had said the quiet part out loud. *Nature* papers are fodder for bragging rights; the more you had the better, and whether the findings were valid or not was immaterial.

When the promised criticisms did arrive in the press, the authors were charged with eyeballing the data and splitting it into two at a seemingly arbitrary point. Vijg then floated a less happy defense: "It boggles my mind how people can come up with these stupid arguments. You see there's a plateau. . . . Your eyes don't lie."[22] Hardly the greatest mathematical proof ever stated but apparently good enough for a top science journal.

If this sounds like grumpy old men fighting over a fishing spot, well yes, but at the time it also seemed to me the sign of a functioning role of criticism in science. The paper was extremely poor, and I assured myself it would be retracted. The dialogue was perhaps a bit unfriendly, but science is supposed to be self-correcting, and it seemed like self-correction was happening. The paper was just an anomaly.

This optimistic assessment was naive. As of 2025, the paper in *Nature*, where zero equals one, is still published, still uncorrected, and still available as an imaginary truth for anyone who wants it. By the metrics that scientists use for career advancement, however, the paper remains a huge success, and the authors have been rewarded with grants and job security that most junior scientists can only imagine.

If you have powerful mates, it seems zero can be one in modern science.

At that time I was unaware of such harsh realities, or at least in denial that they applied to my career. Instead, safe in my obscurity, I had become genuinely interested in the problems of the limits of human life, and I was laboring honestly at getting somewhere in the field.

Early on in these efforts I conducted a thought experiment.

What would happen, statistically, if I faked my age?

Let's say I was aged 40, and I faked my age so successfully that I became aged 50 on paper and nobody could tell. To make a decent sample, let's say there were 99 other age-faking people who managed to do the same thing, and all one hundred of us "young liars" were hiding ourselves in a population of 100,000 real 50-year-olds. This would be a 0.1 percent fake-age error rate.

Assuming we were quite good at identity theft and were not caught, a curious thing would happen to us over time. We 40-year-old "young liars," with our fake ages, are biologically 10 years younger than all the real 50-year-olds. This means that we would be less than half as likely to die than the people with real, accurately recorded ages. So, while we would start out as 0.1 percent of the population, the percentage of people with fake ages would grow roughly exponentially over time. By age 66, fake young liars would be about 0.4 percent of the surviving population. By age 80, the young liars would be about 3.2 percent of the population. Eventually, by age 105, *everyone* would be fake. I would be standing around with my fake-aged buddies, the ones who survived: all of us aged 95 in reality but aged 105 on paper.

This thought experiment is key to understanding how, given time, even a tiny initial error rate in age documents can grow to overtake the entire population. Through a relentless and simple mathematical process—where incorrect data are less likely to die than real data—you would end up in a situation where every single super-old age record is fake, even if you started with clerical age errors of one in a million. This very counterintuitive result means that random errors, even a *tiny* number of errors, could not be ignored when estimating the limits of human life.[23]

In statistics, if mistakes are random, they usually don't affect anything. Here I had instead discovered a process where vanishingly rare, purely random errors had disturbingly large effects on survival models. These tiny errors will especially affect how we estimate the limits of survival.

I was very slow to realize how important this effect may be. Survival models are a key part of how we do medical science. If, say, we wish to know how bad smoking is for you, we use a survival model. If we want to estimate how effective a drug is at preventing cancer, we use a survival model. We even use them to estimate the longevity of cars, fighter jets, software, and phone calls.

Perhaps more disturbingly, the tail end of life beyond age eighty, where this undiscovered effect causes errors to accumulate, has an outsized importance in global planning. Our mathematical models of the future are built by observing the survival patterns of the oldest people alive today, under the reassuring assumption that we will see continuing improvements in human survival tomorrow. The ages where statistical errors silently accumulate—over age eighty—therefore dramatically affects the mathematical models that we use to plan our future healthcare, calculate life insurance premiums, and set the pension rate that decides trillions of dollars in global investment.

Back in 2017, however, the broad scope of this problem was yet to dawn on me. I was only interested in how this pattern of errors affected our calculation of the limits of human life.

As such, in the aftermath of the zero-is-one *Nature* paper, I had begun to write a little theoretical paper about how the biased survival of errors would affect late-life survival in humans.[24] I found that error rates accumulated with age and caused something known as a "late-life plateau."[25] A late-life plateau is the idea that at some extreme age of life, survival rates stop becoming relentlessly worse and flatten out. It has a long history in demography, and, if you believe it is not caused by errors, the late-life plateau means that there is no upper limit to human life. Instead, you end up in a kind of probability-driven game of survival: like playing a non-optional game of Russian roulette every few months. There is no upper limit on your life past that you could not possibly live—you might get a random run of not-shooting-yourself-in-the-head luck—but there is instead an open-ended random process that eventually kills you off.

This sunny view of late life was theorized and promoted relentlessly by Professor Jim Vaupel, the man who crafted the cow-shoveling insult above.[26]

Explaining the late-life plateau as the result of random errors is very problematic for anyone like Professor Vaupel, who spent a long career promoting the idea that plateaus are real and that an upper limit of lifespan therefore does not exist. Yet that is exactly what I did, showing that a simple linear model containing the rate of missing birth and death certificates predicts and eliminates late-life plateaus.[27] I even got my discovery out to review at a very nice journal.

That is why, with a perfect sense of irony, Professor Vaupel and his coauthors published an exhaustive and apparently error-free description of a late-life plateau while my little theoretical paper was under review.[28] This paper dropped a bomb on my life, not because it was a devastating rebuttal of my work but because it was immediately clear that this paper was worse than the original article in *Nature*.

This time the scientific discovery fell in a journal called *Science*. The finding was arguably just as big a deal as the *Nature* paper, and it was making precisely the opposite argument. Barbi et al., as the authors are collectively called, had observed every 105-plus-year-old in Italy and found that there was no upper limit to human life.[29] Instead, they stated that death rates were "essentially constant beyond age 105," and the death by statistical roulette situation of a late-life plateau was therefore real.[30] This was the opposite of what the *Nature* paper had claimed, and an obvious challenge. They could not both be correct.

In a move that neatly infuriated both sides of an angry debate, I realized that *neither* of them was correct.

The problem hinged on the mathematical models. Essentially, Barbi et al. had needed to compare the flattening of late-life mortality to earlier-life mortality rates, to make sure any flattening out was not just random noise at the tail end of their data. So, they had fit a model to what they described as "midlife" mortality data over ages sixty-five to eighty.[31]

Weird, I thought. Since when is midlife defined as age sixty-five to eighty?

Suspicious, I recoded their entire analysis, reran their models, and reproduced their entire results verbatim in a day. Not a single number was out of place. Then, I changed one single number in their equations: I made "midlife" slightly more reasonable and fit their model over ages sixty to eighty. In an instant "the best evidence to date for the existence of extreme-age mortality plateaus" disappeared.[32] The model skewered the late-life data like a shish kebab, and there was no "flattening" at all.

I stared at the screen for twenty minutes and then carefully checked my work. Then, in increasing disbelief, I ran every single combination of ages possible for "midlife": age fifty to sixty-five, age forty-five to eighty, age forty-nine to seventy-three. Without changing their model or data, the existence of a late-life plateau disappeared with *literally every other choice* of ages. Out of the 861 possible combinations with a testable amount of data, their model came out as the single most biased, the single worst fit to the data, and the only model that made a plateau. Without this (let's say) peculiar choice, they would have no scientific findings at all. Only by using the empirically worst, empirically most biased model, which they weirdly called the "best," did they succeed in getting a *Science* paper.[33]

This was either an extraordinary coincidence or a deliberate scientific fraud.

Having learned my lesson about being too slow, I wrote the code in three days, wrote the paper in a weekend, and sent these findings to *Science*.[34] They promptly told me I was free to submit my concerns to an online forum that would not show up on the front of the webpage, was not indexed, was not peer-reviewed, and could be deleted at their whim. Inadvisedly, given the dictatorial status of journal editors, I took them up on the offer and published a salty response in my allotted backroom blog, where I informed them that this format was degrading to the value of scientific criticism.[35] I probably should have stopped, but I was too furious for self-preservation. At academic warp speed, in about three months, I published my takedown in a different journal.[36]

My rebuttal was welcomed like a hedgehog in a balloon factory. Nobody was *quite* willing to crush me to death, given my prickly petite stature, but some of them were ready to whack me with a broom to try and make me leave.[37] After discussion with the editor at my journal, Barbi et al. were allowed to present their rebuttal to my rebuttal: a wholly decent way to clear up an honest misunderstanding. After all, while the odds were 861:1 against it—less than the odds of tossing eight coins and having them all land on heads—maybe it was all just a statistically unlikely coincidence after all. If they had committed fraud, pretending it was a huge coincidence and retracting the paper would also let them leave with some tenuous dignity.

When the time came, they took neither option. Instead, Barbi et al. retracted nothing and suggested it was all a figment of my rich imagination.[38]

Clearly absent from this response was any sense of shame or any whiff of an answer.

Rather than deal with the "coincidence," the authors seem to have either deliberately or (with some convenience) accidentally misunderstood my questions. In the first half of the paper, I had explained that their late-life plateau never existed, and their models were a fluke. In the second half I had explained how, even if it had existed, rare errors could have caused this supposed plateau. They seized on this second-half discussion and pretended this was the only problem raised, stating that my objections depended on "implausibly high error rates for extreme ages."[39]

The results didn't depend on error rates at all: a fact I had told them in private correspondence already, and would later tell them to their face, to absolutely no effect. Yet it hardly mattered. They had ignored the question and replied with an answer to a question I had not asked. Once again that was apparently good enough to prevent any further queries or consequences.

As the drama slowly unfolded like a disused paper napkin, I was starting to suspect that the anomaly of the Dong et al. *Nature* paper was not so anomalous after all. Once again, a zero was a one, and nobody was held to account.

What was *happening* in this field?

2 Saecular Games and the Limits of Life

The previous chapter unavoidably sounds like the detailed griping of a high school drama. It does, however, set up a vital step in understanding the current tone and logic in aging science. These two problematic papers neatly represent the opposing halves of an ancient debate on human longevity: there is either a single upper limit to human life—an age beyond which it is not possible to survive—or there is no definite upper limit to human life and late-life survival is down to chance.

Proponents of the two views have been squabbling like seagulls on cocaine for decades. For an outsider, this has turned every discussion toxic. Amidst the symmetrically accurate name-calling, it also made it unclear which camp was correct.

Fortunately, I was not engaged in aging science just to tattle on grifters and spread gossip. I was also keen to solve problems. Two important questions had popped up: which idea was correct, and therefore what was the upper limit of human survival.

In the hoo-hah over the *Nature* paper, I turned some thought toward these questions and, with substantial good fortune, discovered something new. I found a novel solution to the limits of human life.

First, we need some introduction to aging.

Imagine I have a mouse. If I hit my mouse with a hammer, it will die. Of course, I will then know the immediate or "extrinsic" cause of death of that mouse—the hammer. But if I do not hit the mouse, it will live. For how long? And what will be the cause of death? Well, after a few months or years, my mouse will die without any intervention. It will die from an intrinsic change: it will *age*. The cause of this change, aging, is one that no scientist yet understands.

To begin comprehending why this problem is so difficult to solve, let us conduct another simple thought experiment. Imagine that I have ten thousand baby mice. Imagine once again that I do *not* hit the mice with a hammer, like a normal person. Adequately fed and watered, the mice will live for a few days, months, for two or three years, and then eventually they will all die.

Apart from horrifying my mother and delighting my kids, this enormous cohort of baby mice would allow me to calculate a rate of mouse aging. How? By counting a tally of how many mice are still alive each day, week, or month, I can estimate the probability of survival at each consecutive age. If a tenth of the mice, a thousand of them, die on the first day? The odds of death are about 10 percent at age one day, leaving nine thousand living mice. However, only ninety mice die on day two. The odds of death have fallen to 1 percent in a day.

This rather simple, and onerous, observational experiment reveals a pattern of life common to many mammals. At first, the expected happens—a lot of mouse babies die. For many mammals, including mice, birth is the most dangerous day of your life: mouse mortality rates peak on their first day after birth. After running the post-birth gauntlet, mouse deaths then begin falling dramatically, with the risk of dying halving at a constant rate like radioactive decay, until a kind of brief prepuberty mouse childhood is reached. This initial tailing-off of death rates, from infancy onward, also happens in humans. For example, in rural Gambia roughly one in five infant humans die in their first year (c. 2018 data).[1] Almost all these childhood deaths are concentrated in the first month of life, and a large percentage of *these* deaths occur within the first day.

After the dangers of infancy, childhood is a happy utopia of low-risk calm. Then, after the pleasantries of childhood, puberty promptly attacks. If I keep up my mouse tally, I will find that the odds of any one mouse dying begins to increase from mouse puberty onward, from maturity until the death of the last mouse. When the mice are ten months old, for example, the odds of dying each day may be around 1 in 10,000. By the time they reach twenty months old, the odds of dying each day may have increased to 1 in 1,000. If I keep up my tally from the start of my experiment until the death of the last mouse, I would have constructed a "survivorship curve," a curve that captures the day-to-day odds of death, and incidentally capture the rate at which my mice decay with age.

This survivorship curve has some interesting properties. Let's look at humans, which have a typical mammal-like survivorship curve.

It is rare that anyone includes pre-birth statistics, so let's start with that (figure 2.1). After conception, fetuses have an extraordinary risk of nonviability for the first trimester, though we know very, very little about the earliest stages of pregnancy. It seems that the odds are around 30 percent but fall dramatically each day until the end of the first trimester. Then, the odds of pregnancy loss increase: at first linearly, during the second trimester, then the risk increases at an accelerating rate during the third trimester until the day of birth, when the risk of termination reaches a sharp peak.

After the spike of mortality risk on the day of birth, the day-to-day probability of dying decays extremely rapidly, reaching a safe low point at around age five (figure 2.2). The tally then stays steady. Throughout childhood, all the way to the beginning of puberty, the odds of dying remain remarkably low and relatively constant. For example, in the United Kingdom, there are only around a hundred deaths a year in the age 5–9 and 10–14 category from any cause among a nation of 65 million people: an annual child mortality rate of roughly 1 in 50,000. This pattern is common to all human populations: even in wars or famines, childhood is the safest time of life (figure 2.2).

Then, after puberty, the odds of dying relentlessly accelerate. After your sex hormones kick in, life starts becoming intrinsically more dangerous. There is a brief period until age thirty or so where accidents cause the "accident hump" in mortality rates—usually bigger for men than women. After the hump, life gets more dangerous, at a clockwork and regular rate (figure 2.2).

This clockwork rate is the "mortality rate doubling time" (MRDT) and seems to be constant within a population. For example, if at age forty-eight, your odds of dying are one in one hundred and you have an MRDT of eight years, your odds of death will have doubled to two in one hundred at age fifty-six. They will continue doubling every eight years of age, like this:

Age	Odds of death
56	2 in 100
64	4 in 100
72	8 in 100
80	16 in 100
88	32 in 100
96	64 in 100
104	128 in 100

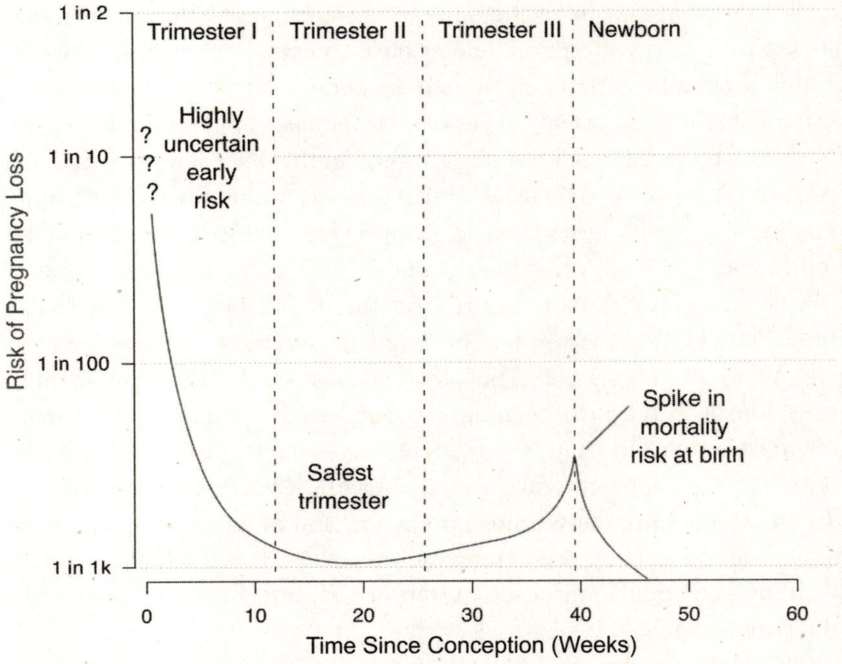

Figure 2.1
Generalized human pattern of pre-birth pregnancy loss and immediate post-birth mortality risk. Measuring the time since conception reveals a conserved risk profile. The greatest uncertainty occurs in the first few weeks after fertilization, especially before the fertilized ova implants onto the uterine wall, but nonviability is certainly very high and highly random. The risk of pregnancy loss falls dramatically after the early weeks post-fertilization, to a low in the second trimester and a relatively safe third trimester. However, mortality risk increases linearly and then nonlinearly as birth approaches, reaching a peak in risk for the developed infant (and mother) on the day of birth. Mortality risk falls dramatically for the newborn child after birth, declining enormously within the first day and month after birth, and continuing to fall at a rapid rate toward childhood. Numbers given here are a loose approximation of risk, not an exact measure. Source: Author.

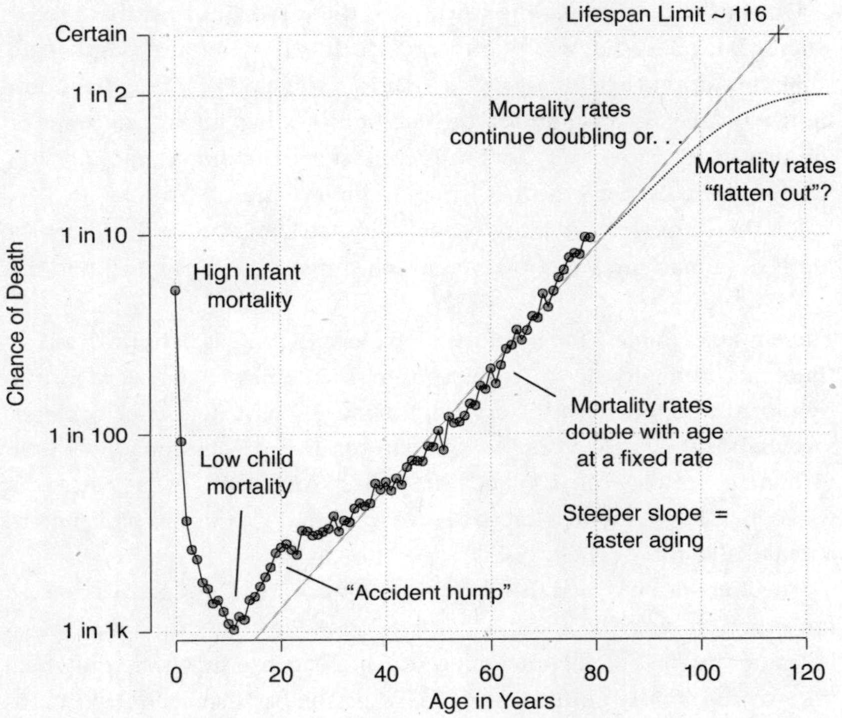

Figure 2.2

The typical pattern of human mortality after birth. All human populations share a generalized pattern of mortality risk (shown per year). After a huge peak in mortality risk at birth, the risk of death declines very rapidly toward a lifetime low during childhood. Mortality risk then remains low until the onset of puberty, when it starts to accelerate. At first, this increase in mortality risk is helped by a jump in accidental deaths, as well as "extrinsic" deaths like warfare and murders, during ages 15–35. This "accident hump" of extra deaths is highly variable in size. Underneath this hump, and continuing for the rest of one's life, is a relentless log-linear increase in mortality risk with age—shown by the gray diagonal line. For adults, mortality rates double every 7–9 years of age. This is our best measure of the aging rate. If we extend the gray line by assuming that the mortality rate keeps doubling with age, we eventually reach an age with a 100 percent risk of death. This is the simplest way to project the limits of human life: here, the limit of survival is about 116 years. However, alternative models make other assumptions—like the dotted "late-life plateau" model that assumes mortality rates flatten out—and can never intersect with the "certain death" line. Data are from Australians born in 1925, but I could have substituted virtually any other population in any other period (e.g., see the hunter-gatherers in figure 2.3). Source: Author.

Obviously, if the doubling carries on, there is a limit* to this process where your odds of dying hit a theoretical 100/100: in our example here, sometime around age ninety-eight. That is, if the mortality rate keeps doubling with age as you approach the end of life—a big "if" that depends on our data at advanced ages—you will reach an age that you cannot possibly survive beyond. This is the mathematical limit of life.

So, the clockwork doubling of mortality rates might help us know the limits of human survival. However, we must first explore our assumptions.

There is an extremely long history to the idea that life is definite, fixed, or finite. According to the writer Censorinus, the Romans had a word for this hypothetical upper limit to survival: the saeculum. A Roman or "natural" saeculum is the longest time any human can live, an absolute upper limit to humans, of between 100 and 110 years.[2] Considering what passed for wisdom at the time, and what passes for wisdom now, this has proven to be a remarkably robust guess.

The Romans borrowed this definition of the natural saeculum as a fixed upper limit to survival, and the idea lives on as the adjective in the "saecular games": games of such magnificence and expense they were only held once *per saeculum* to ensure that no living person had ever seen anything as magnificent.[†] The natural saeculum also served a practical legal purpose, as the Romans were a welfare state and would pay citizens until death. Failing to register a death was, as it is now, a great way to defraud the state and very difficult to detect—how, after all, do you definitively prove someone is dead and not just elsewhere? Therefore, the Romans enshrined the saeculum as a legal cap on survival. It didn't matter if you stood in front of the judge claiming Livia was still alive, if she was over age 110, she was dead.

However, the word saeculum also had an older meaning adopted from the Etruscans—which the Romans call the civil saeculum—that was far more interesting and nuanced.[3]

---

*I called this the maximum survivable age to discriminate it from other "survival limits."

[†]Ironically, the Romans reframed the length of the natural saeculum to 100 solar years, before Emperor Claudius reframed the definition again to 110 solar years so he could move the timing of the saecular games, which cost a pile of denarii but held a lot of propaganda value. Basically, the Romans never stopped muddling with the definition while remembering the pragmatic Etruscan clarity.

The civil saeculum was defined locally by a kind of pattern or algorithm. After creating a new town or city, the first civil saeculum was counted as the amount of time it took between the town's founding and the last death of the oldest town-founding citizen. The Romans wrote down every child born in that year—the first year in which nobody had been alive at the founding of the town—and waited for the last of these babies to die. At this point, a second civil saeculum had passed. From then on, this pattern repeated itself. Another group of babies (a cohort) would be born in the changeover year, they would all grow up and eventually the last would die, and another civil saeculum would have passed in the town.

For example, imagine I build a town called Mundoo-ran. I lay the foundation stones of Mundoo-ran in 100 AD, and a thousand people move to town. After the last of these initial one thousand colonists die, let's say in 175 AD, one civil saeculum has passed. Mundoo-ran is now a civil saeculum old. In 175 AD, a hundred young babies are born in Mundoo-ran. The last of these babies eventually dies in the winter of 260 AD. Mundoo-ran has now, after 160 years, become two civil saecula old. The next crop of babies born in 260 AD will eventually whittle down to zero, at which time the town will become three civil saecula old, and so on.

The difference between the local civil saeculum and a global, species-wide natural saeculum makes clear a useful distinction. Under the Roman intellectual landscape, the local limits of life may change over time and space. Every town has an oldest citizen. Every cohort has a longest-lived member, from the cot to the grave. Yet obviously, this number changes depending on the time and the place of each birth. It is not informative, necessarily, of any global upper limit to human life.

If the oldest citizen in Mundoo-ran, Seattle, or Bremerhaven lives to 103, this does not *necessarily* tell us anything about the ultimate upper limits of human life. It only tells us that the limit must be between 103 years and infinity years old. Narrowing things down to "103, infinity (no limit), or any of the infinite number of possible answers in between" hardly solves anything.

To clarify the problem further, I'll use a simple and fun analogy. Imagine what happens if we replace "lifespan" with "wealth" in the above argument. In the same way I could hypothesize an upper limit to human survival, I might hypothesize a natural upper limit to human wealth. I may tell myself there is some value, some upper limit, capping how wealthy it is possible to be.

How would I measure this value? More fundamentally, how could I test if such an upper limit even exists?

I could, as many scientists do for lifespan, simply observe maximum values.[4] I might take a sample of a population and ask who the richest person is. As above, this gives me some idea, some minimum value to my wealth estimate. If Billy-Joe Goldtooth is the richest person in town and they stroll about with $8 and a banjo, I can confidently surmise that the limit of human wealth is at least $8 and a four-string.

But what does that tell me about all of humanity? I only looked at one town, after all. I clearly need to expand the search in case my hometown is somehow atypical.

I could ride to the next town and retake my sample, and the next, and the next. This allows me to observe, but it also allows me to generalize. For example, I might decide to visit a hundred random towns. Along the way I might find that for every twenty Billy-Joe Goldtooths who own $8 and a banjo, there is exactly one feller named Hairy Buckeye Holewillie who owns $9 and a mule: 50 percent more wealth than Billy-Joe possesses. Then, I might discover that for every twenty Hairy Buckeye Holewillies out there, there is exactly one Jenny Woodchuck with $9.50 and a whiskey still: this is 25 percent more wealth than H. B. Holewillie.

This sampling has allowed me to discover a mathematical relationship between wealth and population number. As the population number increases twentyfold, the upper value of wealth increases by a little less each time: we have found that wealth is asymptotic to population size. Through some simple mathematics, the limit of human wealth can therefore be estimated at about $10, a shotgun, and a three-legged dog. Of course, such unimaginable richness will never be achieved: this is my projected, unreachable limit of human wealth.

Many scientists undertake a similar process when they estimate lifespan limits. They sample populations across a map, discover some relationship between population size and the upper value of longevity, and use these data to estimate an upper limit of human life. Indeed, this appears to be the main method for calculating the upper limits of human survival. This is the kind of approach I used above with the MRDT: in most populations, the risk of death doubles every 7.5–8.5 years with frightening regularity, allowing the estimation of a limit to survival. Add up enough towns, you may think, and these estimates will converge on the true value of the "maximum survivable age" beyond which humans cannot live.

However, there is a critical flaw in this thinking.

Let's reuse our wealthiness example. Imagine I get on my mule and ride to Billy-Joe Goldtooth's hometown of Black Lake. There I find out that Billy-Joe's older sister, Smelly Jane Goldtooth, has $50 million and *two banjos*. My initial wealth model falls apart. Everyone in Black Lake is rich: I have discovered, to my horror, that my relationship between wealth and population size is not universal. I assumed that the pattern of wealth was the same in every town before I could observe every town. In statistical terms, I have overfit my model to a population subset and assumed it is representative of all humanity.

If I discover my assumptions are wrong, and the mathematical relationship between population size and wealth changes when I reach a new place, my calculated limits become illusory and useless. I can no longer extrapolate a global limit to wealth from my observation of typical hillbillies. Instead, wealth is "nonstationary": the pattern of wealth changes nonlinearly as I walk or ride across the map. Some places are wealthy, or long-lived, and some places are not. Worse, some places may look alike (the countryside), while some are outliers (Black Lake or Big City). If I look at some populations, it does not necessarily inform me of all the others. There could always be a Sedona, a Clark County, or the Douglas Ewing Ranch: a statistical oddity hiding in the weeds, waiting to upset my nice little mathematical models.

This problem does not only apply to changes observed across the map but also to those over time. For example, imagine that Billy-Joe Goldtooth falls over, his banjo pokes into the ground, and he strikes oil. The entire town becomes rich. Suddenly, because of some unforeseen change in the environment, the distribution of wealth changes dramatically. The old constraints on wealth no longer apply, and the apparent limits of human wealth triple overnight.

The parallel here with lifespan is hopefully obvious. Many scientists have attempted to calculate the upper limit of lifespan by observing extreme lifespans in as many places as possible. If we know the oldest person in the world, the thinking goes, we can estimate the oldest it is *possible* to be. However, as our wealth-limit example shows, there are deep flaws with this idea. Just because Crassus was the richest Roman does not mean it is impossible to be richer than Crassus—just ask Rockefeller, Mansa Musa, or Atahualpa.

We could assume that the maximum lifespan we observe in ten thousand Roman towns tells us about the upper limit of all human lifespans. We may also be more tricky and estimate the limit of survival in every Roman town. But to be sure that these methods work, we need to know how (and if) these estimates change as we observe new places and new times. Gathering more data should just help us zero in on a more accurate estimate: if I take larger and larger samples, my estimate should converge toward the real value.

Unfortunately, of course, I have no fundamental reason to assume that this will ever be true. In fact, I can apply some clever mathematics to the towns I do observe and solve an equation that shows such logic is definitely *not* true.

Much like the distribution of human wealth, the extreme values of the human lifespan are not a stable, stationary distribution. Rather than narrowing down to a single value as I measure more populations, the upper limits of human life experience endless change.[5] The same thing happens when I estimate the limits of survival using the MRDT or many other mortality models.

The limit of human life does not converge closer and closer to one maximum value over time or space. It does not converge as the population gets larger and the data become cleaner.[6] Instead, the larger the population observed, the more the upper limit of lifespan jumps about. In mathematical terms, the data are called "nonstationary."[7] This means that even if we accurately discover the oldest human on earth right now, this does not necessarily tell us the limit of human life, any more than peering at current billionaires tells us the limit of human wealth.

Even if we observe every person, for wealth or lifespan, it does not guarantee us any information about the past or future value of such limits. More remarkably, we *know* that we cannot forecast these values, for the same reason we know we cannot make a six-month weather forecast, because these values are intrinsically unpredictable.[8]

We may say that Bill Gates is the world's richest man with umpteen billion dollars, or that Jeanne Calment is the world's oldest woman at 122.5 years. Yet, if true, these values only give us a minimum: they tell us that any prospective limit to human wealth or life must be between this value and infinity. Beyond these basic constraints, however, these maximum data cannot tell us our limits or if any limit *even exists*.[9]

This finding provides us with something brand new. The evidence that human populations do not converge on a single limit to survival, and that we can demonstrate this non-convergence mathematically,* allows us to throw out the idea of a single limit to human survival.

I realized all of this and set about understanding what it meant as I labored on the rebuttal to the deeply flawed *Nature* paper.

But what of the other answer? To recap, this is the idea that the odds of dying flatten out in extreme old age so that the ongoing chances of survival resemble a thrice-yearly game of Russian roulette. There is no upper limit to how long you can live in this scenario, just a constant chance of death that gets you in the end. This is called a late-life plateau. Not a rosy picture of old age.

I had already infuriated the people who believe the plateau is real by pointing out two things. First, you can generate late-life plateaus by committing a tiny number of random errors. Second, most of the data suggesting that human life is unlimited can be easily explained as an error pattern, with over 80 percent of the variation in late-life plateaus explained by error-generating factors like "the number of people with birth and death certificates."[10]

I am about to infuriate them further with the words of our "cow-shoveling" Professor, Jim Vaupel. Professor Vaupel came up with the best, and by many means only, explanation for the late-life plateau. In 1979, he used data from Swedish women born in 1875 to estimate that the plateau began at age eighty-five. This discovery made his career and has been quoted in thousands of scientific papers.[11] His explanation was a biological one: the plateau was caused by natural variability in the people he was measuring.

Curiously, as Professor Vaupel got older, and the data he used got progressively cleaner and better, the plateau moved itself. This even happened when the plateau was being reestimated in the same populations. In 1979

---

*By using the Kwiatowski-Phillipps-Schmidt-Shin (KPSS) test for nonstationarity, for example, to rule out convergence to a unit root (and rule out trend stationarity) for the human lifespan limit over time. Easy-peasy, failure of the null at $p < 0.00001$. Or, if you like, the Dickey-Fuller test for nontrending unit roots. Whatever takes your fancy, really.

and 1990, Swedish women had a mortality plateau at age 85, but according to Vaupel, the exact same Swedish cohorts in 2017 "suggest the appearance of a plateau at about 103 years of age for men and somewhat later, around 107 years, for women."[12] Even this conclusion is flimsy: plotting the raw cohort data directly from the Gerontology Research Group (GRG) database produces a ramrod-straight acceleration with age, not a plateau, with no deceleration at age 85 or anywhere else.

Why the same plateau, estimated from the same data had retreated twenty-two years, during forty years of data cleaning, seems to be neither asked nor answered. To me it looked suspiciously like the plateau was running away to progressively older ages, retreating to escape from the cleaner data of modernity. This pattern also happened to everyone else's plateau data, including the Barbi et al. team with the coincidental choices who moved it to "after age 105."[13] Strange behavior for a supposedly fixed and universal biological pattern, but perfectly reasonable if you assume that the plateau is caused by errors.

This ongoing shift in the plateau's age also neatly throws shade at the other biological explanations for the plateau. A real, robust biological effect caused by evolution would not disappear when your data gets cleaner.[14] Yet that seems to be exactly what has happened.

Even more bizarrely, the natural variability used to explain this plateau, called "hidden" heterogeneity or population heterogeneity, is not and has never been measurable.[15] The starting idea of hidden heterogeneity is that certain people are more or less likely to die than others due to unmeasured and unknown factors, which is reasonable by itself, and that the right amount of heterogeneity can cause a plateau. But how variable is this interpersonal risk? Well, it's hidden by definition—purely theoretical. As a result, the degree of hidden heterogeneity is, like Einstein's cosmological constant, modeled without ever being measured in a process known to the public as "making things up to fill a gap."[16] This raises some difficult questions, like how did anyone derive a numeric value of heterogeneity to build their demographic models, or how can they pretend to know it is a good-fit explanation for observed data, for the last four decades.

So, while we cannot definitively rule out the "no limits" answer for now, we can ask some awkward questions about why the changing value of the plateau has been ignored for four decades and set the whole idea aside as unsupported.

As I worked my way through this logical muddle, it put me in an apparent conundrum. The idea that there is one limit to human life is mathematically ruled out: every time we ask the question, we get a different answer, and these answers do not converge on any one number. Yet the evidence for a "no limits" solution is, in my opinion, extraordinarily bad and ever-changing.*

Neither the argument that there is one limit to human life nor the argument that there is no limit are well supported by the data. The answer to the first question we asked—"which of the two answers is correct?"—was apparently "neither of them."† After all, if one side were simply correct, decades of modern debate should have settled the issue. Perhaps it makes sense that a debate can only survive so long if neither side is right.

However, I promised a solution to the question "how long can a human live?" Here we go.

I can use my sampled data on lifespans to mathematically extrapolate a limit to human life. I can also, given enough data, extrapolate a limit to monkey life. Yellow baboons, to be specific. And I have, shown below (figure 2.3). The yellow baboon has a limit of survival of about thirty-one years.

So, what can I conclude here?

It is uncontroversial to state that the monkey has a different estimated limit to life than the human. But if the limits of lifespan vary between a species (figure 2.3), why not *within* a species?

What if the Roman/Etruscan idea, that every town has an ever-changing and unique limit of life, reflects a biological reality?

This was my answer to the long debate on human lifespan limits. Not that there is one global limit to life, or no limit at all: there are *many* limits to human life that change rapidly and unpredictably‡ over time and space.§

---

*I had not thrown out the evidence yet, which would come later, but I had very strong doubts.

†While I am sweeping the logic under the rug for now, in the next few chapters I will pull apart why the "plateau" model does not make sense.

‡Or, more correctly, in a nonstationary way that limits predictability. Like the weather, nonstationary processes can often be predicted into the future for a short time but not a long time.

§A much less preferred answer is that if the "no limit" people are correct, and a plateau existed, then any such plateau reached would be highly variable across species and even within humans.

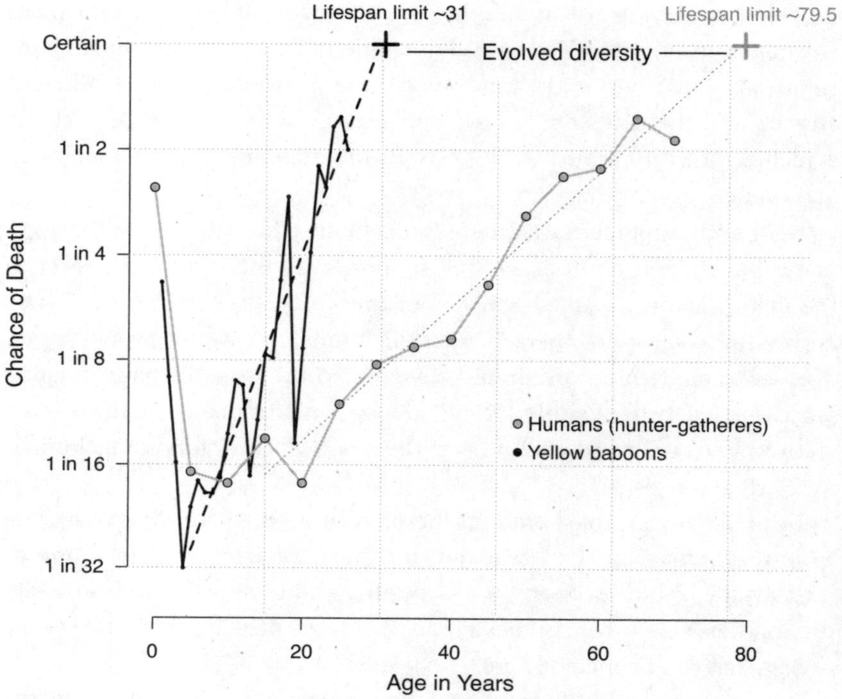

Figure 2.3
Lifespan limits in hunter-gathering humans and yellow baboons (*Papio cynocephalus*). Lifespan limits are obviously different between species, as shown here, for example, between human Agta hunter-gatherers from the Philippines (gray lines and points) and yellow baboons (*Papio cynocephalus*, noted by black lines and points; mortality rates are shown per year for baboons and per five-year period for humans to make things visually easier). Humans and baboons evolved from a common ancestor to reach very different lifespan limits—here estimated as thirty-one years for the baboons and almost eighty years for the humans (much lower than our previous estimate). There is obviously between-species diversity in lifespan limits, or between-species diversity in late-life plateaus if you believe the alternative. But for evolution to act and for natural selection to take place, there must also be the capacity for within-species diversity in lifespan limits (or plateaus). That is, there is no reason to assume only one limit to human life. In fact, we might expect a huge diversity of lifespan limits: in the same way that humans have a huge diversity of skin color, height, or lung capacity, each affected by some combination of genes and the environment. So, what sense does the search for a single limit even make? Why not suggest the third solution: not one limit, or one plateau, but the potential for (modifiable) human diversity in each. Source: Author.

This is a far, far more exciting answer. Because the rates of change observed are so fast, occurring over a matter of decades, we can reasonably rule out genetics as a cause of such diversity. Genetics and gene frequencies takes eons to change.

Instead, the limits of human survival must depend on hidden factors in our environment. There are many limits to human life, and these limits are changing stupendously fast in response to unknown environmental causes. Because we have never measured the diversity of human lifespan limits before, we have no idea what these factors may be.

The question "how long is it possible for a human to live?" has a new answer: the mathematical limit of survival changes very quickly over time and depends on factors in your environment, and we don't know what these factors are. Now, for example, the limits of human life are rapidly increasing across the globe after cratering in the 1950s. This is even evident in specific populations: in Sweden in 1775, the estimated limit of human life was around 120 years.[17] By 1950, it had fallen to 109 years, but by 2010, it had risen again to about 115 years (figure 2.4).

Why? I have no idea.

I don't have any answers at all. And that is deeply exciting.

Regrettably, few of my fellow scientists saw it that way. After my attempt to have two high-impact papers removed from the record, demography rather resembled a play of *Romeo and Juliet* where a stranger had run on stage, kicked Mercutio in the downstairs jumble, slapped Tybalt across the face with a dead fish, and eloped with Juliet. All sides were furious at me and so were the onlookers.

Fortunately, among the prominent-person hate mail, one comment had lodged itself in my brain. Ken Wachter, the senior author of the Barbi et al. paper, had thrown a rock at me in peer review about my simple theoretical model explaining the limits of survival. If my theory were correct, he said, virtually all extreme-age records—all supercentenarians (people over 110 years old) and almost all centenarians—would be false.

Wachter had meant this as an insult: a statement that I was obviously mentally deficient, or my theory flawed, because of the far-reaching and (to his mind) absurd consequences of the mathematics. I decided to take him seriously. You should never discard the logic of people who dislike you merely because they dislike you. After puzzling it over for perhaps a week,

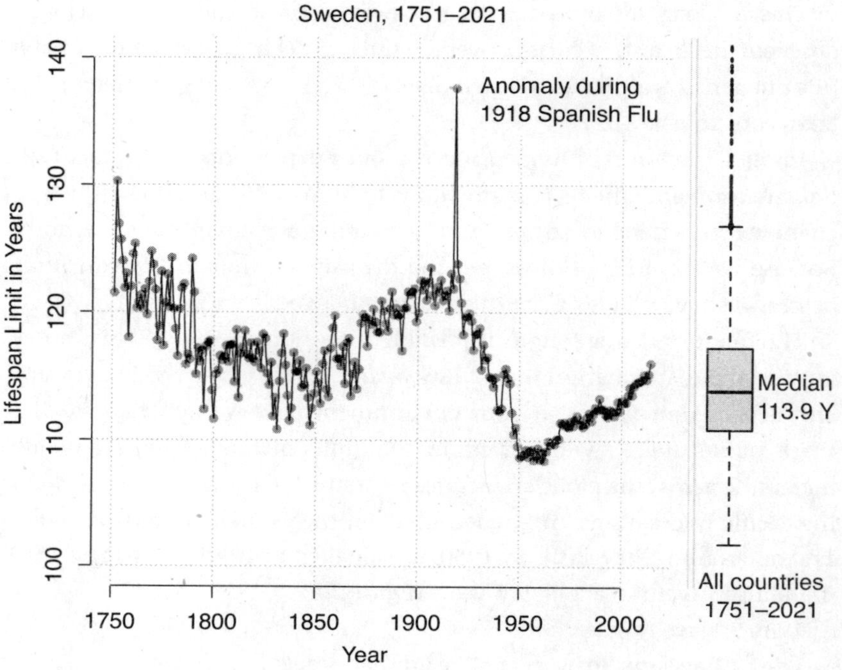

Figure 2.4
The dynamic and diverse limits of survival. In every measured country we see a dynamic, nonstationary change in the projected limits of human survival. Here, the remarkable historical records of Sweden allow more than 270 years' worth of calculations for the maximum limit of survival. Rather than converge to one global limit, Swedish data reveal major, rapid changes in these values over time. Some distortions occur through extrinsic factors, notably during the Spanish Flu, which primarily killed young adults and distorted mortality patterns. (It is also worth noting that Sweden was neutral in World Wars I and II.) Other changes are longer-term trends, but these can reverse or change across very short timescales. As a result, these data have a mathematical property: they are nonstationary and nontrend stationary, meaning they keep unpredictably changing over long timescales. As a result, there seems to be no single limit to human life. Instead, humans display a huge and flexible diversity, across a range at least as large as that shown in the boxplot in the right-hand side of the panel. Source: Author.

I decided that Professor Wachter's logic was justified. He was right. If my mathematical thinking was sound, decades of demographic models would have to be discarded, and our carefully constructed global data on centenarians would be dominated by statistical garbage.[18]

At first, the case against me seemed enormous. Decades of apparently painstaking validation work was used to certify the world's oldest people. The esteemed Professor Vaupel had even used his fame in explaining the late-life plateau to appropriate tens of millions in euros from the German government to found the Max Planck Institute for Demographic Research in Rostock. A central purpose of the institute was creating a database of centenarians—people over age 100—and so-called supercentenarians who had exceeded the remarkable age of 110. There was also a dedicated American group, the GRG, backing these data up and the more general involvement of large numbers of validating scientists.

For my mathematics to be correct, all this work would have to be mistaken to a shocking degree. Worse, if these data were errors, then our understanding of other important questions like "How many humans are alive?," "Does aging ever stop?," and "How many old people will exist in the future?" would be dramatically wrong.[19]

It *seemed* I was wrong. Yet the reaction and apparent cover-up of my scientific peers, their overtly hostile responses, and the striking obviousness of the mistakes that had been both made and overlooked shook me. Could it be true? Could most of the world's oldest people be just . . . shonky data?

With heavy resignation, I loaded up the largest centenarian databases in the world and decided to find out.

3 Dining with Kubrick

It is 1991 and Stanley Kubrick, the bearded Bronx-raised film director of incredible fame, is in London. Julie Walters and Patricia Hayes are starring in a remake of *Rose Tattoo* by the Peter Hall Company, and it is running at the Playhouse Theatre in Charing Cross, London. On a whim, Kubrick attempts to get a seat to see the show. The theater is full, and the usual problems play out until the American director, reluctantly, pulls rank. He quietly states who he is to the box office attendant: he is the director of *Dr. Strangelove, A Clockwork Orange, The Shining, Full Metal Jacket*, and *2001: A Space Odyssey*. He is, in short, the world's most famous director and one of the best-known people on earth.

Predictably, the worker at the box office finds him a seat.

Kubrick duly goes through the show, watches Serafina mend the silk, and graciously agrees to an after-show chat with the actor Julie Walters. After a warm chat with the cast, in which Walters discusses Kubrick's films in detail and Patricia Hayes professes she would love to work for Kubrick in the future, the night concludes with a warm farewell.

This unremarkable and pleasant evening is the beginning of a somewhat consistent return to London for Kubrick who, despite his normal preference for hermitude, begins to slowly become a regular feature in the London social scene. The warm reception he receives from attending the London show is, in his words, "a big buzz."[1] As a result of this one night, while filming *Eyes Wide Shut*—and afterward—Kubrick starts to hobnob in London. He begins consistently engaging with people across the city over the next seven or eight years: meeting old friends with whom he has worked on films, giving semiregular and often impromptu interviews to the press, and encouraging young actors and actresses to dine with him, usually by dangling the promise of future acting roles. He even signs loans and runs up charges, financially supporting causes such as a gay nightclub and dining out on his friend's credit cards while the studio (he promised) would pick up the tab.

The problem was, of course, that the probably best-known film director of all time flitting around London like a New York socialite was not a straight, black-haired, bearded American from the Bronx. He was a beardless, British, openly bisexual alcoholic conman: a terrible parent with a beautiful English accent and gray hair.[2]

For almost eight full years, the London-going "Kubrick" was an impostor, an absurdly mismatched one, originally named Eddie Jablowski. This distinctive chancer made his first known name change after getting out of prison, from Eddie Jablowski to Alan Conn. He later settled on a semipermanent name after realizing his passion for being a conman: he loved the con way of life, so he named himself Alan Conway. Conway had a fun time being Kubrick. In his words, "whenever I had a few Vodka and Tonics in me, Mr. Kubrick was around."[3]

Conway had a striking ability to pass himself off as one of the world's most famous people in one of the largest cities in the world, usually while drunk and often to Kubrick's own long-term colleagues. What is even more remarkable is that the Conway con persisted for years, not days, to the amusement of the real Kubrick and to the consternation of his dupes. With no identification papers, no American accent, no money, no sobriety, and no beard, Conway fooled reporters and friends of Kubrick, right out in the open, just twenty miles from Kubrick's house in Childwickbury. Conway did such a superb job of being Stanley Kubrick that John Malkovich wanted to emulate the feat, pitched and wrote a script, and acted as "Conway pretending to be Kubrick" in a film.

It was fantastic.

The imitation caper only came to an end when Conway was arrested for signing a bad check. He promptly pleaded insanity, faking it all the way, and mastered that role as well. He was sent to a mental health facility, which almost immediately cleared him as mentally stable, and was sent home without further charges. Before his death, he stated "I enjoyed being Kubrick, [a] great part for anybody to have."[4] It also seems to be a part that he needed to escape into, considering a tragic and conflicted real life that included the death of his gay lover from AIDS, heavy alcoholism, and his persistent abuse of his son.[5]

The Kubrick hoax remains a massively unlikely and bizarre identity theft, attested by dozens of witnesses, the perpetrator, the perpetrator's son, the extraordinarily famous and somewhat amused victim Kubrick, Kubrick's wife, Kubrick's lawyer . . . a good portion of the London theater scene, it seems, witnessed the hoodwinking.

The point of recounting such a remarkable story is to highlight it as an incredible and unlikely farce: one that would seem utterly stupid if you merely imagined it happening, without knowing that it *had* happened.

This singular and irreproducible fraud will be relevant, in a rather surprising way, because we need to talk about the oldest people in the world.

The oldest man in history is arguably the second-most critiqued record of human longevity, though the second-oldest woman (by merit of outliving the men) may also lay that claim. His case is, if not the gold standard, probably silver or at least in the Olympic finals for "good evidence in gerontology."

The world's oldest man is validated, as with all "oldest people," by exhaustive examination of his documentary records by professional demographers. Jiroemon Kimura, born in Japan as Kinjiro Miyake, has a reported age of 116 years and 54 days.[6] He had eight children, one wife, and became the world's oldest man after overtaking Christian Mortensen in 2012—whose claimed secret to longevity was smoking "a good cigar."[7] All of which is attested by documents of varying types.

Jiroemon Kimura also has at least two birthdays—an unusual little wrinkle, you would imagine, for a valid world-record holder.[8]

The wrinkle becomes even more wrinkly when you discover that one of Kimura-Miyake's two birthdays was deliberately faked, and he lied about it. Under interview, Miyake's extra birthday was explained in a way that was "not feasible," and it was promptly concluded that this extra birthday had been forged.[9] The birthdays had been copied in 1955, from earlier records that are lost, and we now use birth data that were promoted by his relatives.[10]

I hear you, skeptical reader, say sure, okay, Kimura-Miyake's birth records might be faked. Surely, aside from the deliberately faked birthday—which seems kind of important when claiming to be the oldest man in the world—there must be some other convincing evidence of his age to make up for such a shortfall. Well, let's look at the other evidence.

Kimura-Miyake has three marriages to the same woman, recorded on different dates, yet no apparent divorces.[11] This rather strange practice raised no doubts but rather seems to have been optimistically entered as being triple the evidence of one ordinary marriage. Amazing. Fitting this theme, Kimura-Miyake was also conscripted to the Imperial Japanese Armed Forces three consecutive times.[12] This documentary evidence seems, at least, to not contradict his claims directly. Unless, that is, we tug at a little thread. At the time, the period of mandatory conscription into the Imperial Japanese Armed Forces was a three-year term.[13] Being discharged from military service often resulted in severe beatings or jail and generally meant you could not reenter military service. Exiting legally meant you either had bought your way out, at rates far beyond Kimura-Miyake's means as a telegraph worker, or you carried a medical exemption that excluded you from service forever.

Japanese military manuals allowed summary execution without a trial for conscientious objection, and minor infractions were punished by severe beatings—often to death.[14] Once conscripted, Japanese people crossed a "point of no return."[15] How, then, was Kimura-Miyake able to be conscripted three times in four years, casually leaving and returning to the armed forces for three-week stints? It is, let us say, a little unclear.

In the month before his marriage, his employer the Imperial Japanese Army had massacred and raped its way through northern Korea, killing over five thousand people in one village alone, including by burning people alive under hay bales soaked in gasoline.[16] Where he was physically working, they were erasing Korean history by systematically demolishing ancient Korean palaces to put up barracks: the equivalent of dynamiting the Louvre to make a McDonalds.

Did he get out of this with a kind of hall pass? Did he say, "You know, guys, I'm not really *feeling* it today," and they just let him leave?

Equally unclear is how he was able to legally marry given that marriage was illegal during the first three years after conscription. Odd, given that he did it three times in the middle of his conscription periods.[17] Perhaps the famously easy-going Imperial Japanese Armed Forces of the interwar period—responsible for the murder of millions and the assassination of their own head of state for insufficient patriotism—just shrugged their shoulders, ruffled his hair, and let him off the hook?

Another strikingly unusual part of the Kimura-Miyake case is the name change. In 1896, Act 89 of the Japanese Civil Code established "Article 750" (later also supported by the Family Register Act Article 74, item I), which legislated that married couples must have the same surname at marriage. Kimura-Miyake did not have the same surname as his wife, at any of his three recorded marriages. For example, he had one marriage (apparently the real one) on December 27, 1920, after which he mysteriously waited almost eight years to change his name on April 12, 1928. However, he did not change his name through official documents. No, bizarrely, he just scribbled it on a piece of paper.[18] His family apparently kept this rock-solid evidence—an old scrawl on a Korean telegraph company letterhead saying "I changed my name"—and it now counts as documentary evidence to support his case. It was considered excellent evidence.[19]

The nature of the name change adds to the deep weirdness of the case. Kimura-Miyake took his wife's last name—which was unusual—and simultaneously changed his first name—which was *very* unusual. Even now, less than 3,000 Japanese people a year change their name, overwhelmingly just their last names, in a population of 126 million. Presumably not many of these are adopting doubly new names. In the far less constrained atmosphere of Japan in 2024, compared to the prewar hyperconservative and tradition-obsessed Japan of 1928, an unknown but small minority of husbands adopt their spouses' surname. Some 96 percent of women now lose their surname at marriage, likely limiting the number of husbands adopting their wives' names to below 4 percent. I suspect a good fraction of this 4 percent—about 120 people a year—includes couples adopting mutually novel surnames rather than husbands taking their wives' name: for example, "Mr. Takeda" and "Ms. Honda" becoming the couple "Takeda-Honda."

Why Kimura-Miyake apparently broke the law for years before adopting both his wife's name and a brand-new first name, without registering anything or telling the authorities, is entirely unknown. Worse, nobody seems to have asked why he did it this way. Just another anomaly in the paperwork; nothing to worry about. I mean we all permanently change our names, to our wives' last name and some arbitrary first name, by writing "I change my name" on some work stationery, don't we? It's a perfectly normal and legitimate way to do it, right? Apparently, the scientific community thinks so.

Sharp readers may notice that the remarkable series of evidence on the world's oldest man comes largely from a single source.[20] This could be problematic because maybe I am echoing the work of some cynical critic, highlighting fundamental and absurd cherry-picked flaws in the Kimura-Miyake case. However, you can stop worrying. The multiple-birthdays-and-weddings name-changing calamity of the Kimura-Miyake world record is not sourced from a demographer who wants to derail the case. Instead, it comes from the people who concluded that Kimura-Miyake is telling the truth. Incredibly, I am quoting evidence from the people who crowned Kimura the world's oldest man.

As a final testament to their thoroughness, these scientists report Kimura-Miyake's older brother's birthday as "February 6, 1985," a few days before my own birthday, and for the oldest man in the world, confidently state that "J.K. [Jiroemon Kimura] was born on April 19, 1987."[21] Off by a century, twice. According to demographic experts, however, nothing could get by such brilliant minds, especially not a badly recorded date.

In Japan, where the birth and death certificates are self-reported, and where "dying alone and not being found for ages" is so common that it has its own particular word (*kodokushi* if you want to know), the self-promoted case of a man who lied about his own birthday and made up his own name on a scrap of paper is deemed to be completely bulletproof. No questions, no concerns.

I may go find some East India Company office stationery, rename myself "Viscount Arbury Ponglesmith-Bluebeard the Bachelor," and make myself 206. It seems that easy.

Despite his multiple birthdays, marriages, conscriptions, and names—generated in ways that directly indicate fraud or extensive error—the Kimura-Miyake story has been swallowed up whole by an eager academic community without so much as light indigestion. The public and press have, inevitably, followed them. Kimura-Miyake is a hero, almost universally regarded as the world's oldest man.

Like young and aspiring actors drinking with "Stanley Kubrick" in a SoHo bar, people need to believe the Kimura-Miyake case. And so, they do. If you want it badly enough, you just don't care that Kubrick is beardless, drunk, and British.

Equally credulous but ultimately flawed cases have arisen before and are well-known among some (suspiciously silent) scientists. To my shock, I soon began discovering such cases at an incredible rate.

Disturbingly, most did not require much detective work to find and adhered to an uncomfortable general pattern. An extreme longevity case would be announced, get catapulted into the global press and lauded by everyone, and disappear years later once the evidence was checked. The "delayed quiet failure" part of this process is especially weird: cases announced with fanfare and pomp to the entire world are, when discovered to be statistical nonsense, withdrawn from the record books in monk-like silence.

I found Carrie like this—a forgotten case of a discarded human, used by the scientific community for self-promotion and used by the newspapers for selling copy, before being dumped quietly by the intellectual roadside. The details were beyond bleak. Carrie White had been the world's oldest woman long before Jeanne Calment: for twenty-six years, she was adored, feted, questioned, certified by the Guinness World Records, and validated in the eyes of all demographers using the same methods used today. International newspapers treated her longevity and her birthdays as a major news event. It was then discovered, through a single typo in a long-overlooked document, that it was all nonsense.

A psychiatric hospital orderly had written down her birthdate incorrectly. They had, according to our high-tension acquaintance Robert Young, simply noted her birthday down as the first full day she had entered the hospital minus exactly thirty-five years.[22] All subsequent records of Carrie White copied her birthdate from this guesswork error or from other records that also copied the error. Whenever someone needed to know White's birthdate, they had just copied it over from the made-up date.

This even happened, somehow, when White's earlier paperwork was lost and needed replacing. This kind of error usually arises something like this: "Lost your birth certificate? Don't worry, present the bureau of birth certificates with evidence of your age from your current documents: a driver's license, a bank account, a letter from your lawyer. Let's just copy the birthday on those and issue you a replacement." When someone does this, it becomes possible to propagate an age-coding error backward through time with perfect consistency: all your documents, even the ones issued before the error, will have the same incorrect birthdate. This is essentially what happened in the Carrie White case, but in a mental institution rather than a government bureau.

In a result that reveals how deeply vulnerable the paper records of humanity are to human errors, her incorrect birthday was propagated both forward and backward through time. The outcome is that for twenty-six years, Carrie White was the oldest and then second-oldest woman on earth.

White did not seek to propagate the myth of her longevity herself. She was not even promoted by her relatives, like Kimura-Miyake, because she had none. Instead, she had seemingly lived out her adult life in a refurbished Civil War–era prison known as the Florida State Hospital for the Insane: an institution famous for losing a Supreme Court case after confining a clearly sane man against his will—alongside over one thousand men monitored by just one doctor—for fifteen continuous years.[23] That man, Kenneth Donaldson, described White's psychiatric home as a place of torture, chaos, and electroshock horror.[24] Another patient, Chris Calhoun, revealed that staff would provoke fights between mental patients and then choke the winner and loser alike until they were unconscious. Calhoun saw several people choked to death this way, amid unchecked rapes and violence. A happy healthy home for the oldest woman in the world.

White had reportedly been committed to this nightmare for life after a diagnosis of "post-typhoid psychosis" manifested by her "exposing her arms" (i.e., rolling up her sleeves).[25] At the time, people were routinely admitted to the Chattahoochee asylum for signs of insanity like "being difficult," nude sunbathing, or becoming pregnant to "a full blooded Indian,"[26] so her forced admission by a long-evaporated husband was the standard level of cruelty and lack of care for Chattahoochee.

Given the institutional disorder and the striking number of unmarked graves out back, it is not even clear that Carrie White is Carrie White: she has no contactable relatives, the diagnosis of "post-typhoid psychosis" is sufficiently creative that "doctors now are not certain of [the] diagnosis" because it doesn't seem to be a real disease, and her admission papers note that she was concerned "someone was harming her twins, but she didn't have any twins."[27] When asked about her past, her husband, her children, or her relatives, she simply never answered.

While she was paraded like a P. T. Barnum sideshow for her longevity, by an insane asylum trying to salvage a destroyed reputation, not a single person questioned the Carrie White case. Instead, it became a global phenomenon: her 115th birthday was feted in the *Los Angeles Times* and *The Washington Post*. At 116, Ronald Reagan called her personally for what must have been a substantially odd birthday chat.[28] Her death was discussed and mourned worldwide. Yet despite the abundant investigative reporters, none had asked any basic questions.

How exactly was it that the world's oldest woman had outlived us all, athletes and monks and health gurus alike, while being an inarticulate inmate of a notoriously brutal insane asylum? How were *these* the living conditions that caused world-record-breaking health? Why would we believe the age of a woman who could not remember her own children based on documents kept by a failed institution? Where was her birth certificate or her family?

White's record had come to light in the shadow of an infamous court case that had forced her release, after which the asylum that had generated her (inaccurate) records was desperately attempting to recover some threads of credibility. Yet people simply swallowed it all like a dose of Thorazine.

Indeed, it seems they still do. As I write this in 2024, the world's oldest man hails from Liverpool—one of the poorest parts of the United Kingdom—and he took the crown from Juan Vicente Pérez. Pérez grew up in a notorious cocaine-growing region of rural Venezuela, did not own any identity documents until he was 54 years old, was involved in a pension fraud and identity theft dispute aged 109, and was promoted by the dictator Nicolas Maduro as an indicator of Venezuelan greatness.[29] Venezuela currently has the same life expectancy as midwar Ukraine,[30] without the assistance of Russian rocket bombings, and yet still nobody asks any questions.*

As I stared at case after case, I began to make an uncomfortable realization. What was striking me was not that the frauds and errors were eventually detected but rather how utterly each "validation" had failed to detect a flaw. The cheating happened right in front of the scientific and journalistic community for decades, but like a Lance Armstrong drug test, the fake 110-year-olds kept being cleared. Case after case, time after time, the gymnasts with the weirdly deep voice and the fresh needle-track marks are written down as "validated," as clean and clear as newly melted glass.

---

*Like why is living in a cocaine-rich jungle supposedly good for you, why does he credit his longevity to fish and chips—the culinary equivalent of smoking three packs a day—and also what about that Venezuelan bloke?

Carrie White had passed all the tests that demographic science could deploy. Were it not for the survival of one stray document, she would still be validated.

Something was deeply wrong.

Debunking a case that is widely accepted by the scientific community is like seeing a piece of a movie set fall over: you realize in a moment that the saloon in the spaghetti Western is not a two-story building but a painted two-dimensional prop made of plywood. You have the illusion of solidity in one moment, an incredulous feeling of having been tricked in the next, and after a stunned second you begin looking at the entire street thinking, "I bet that brothel's made of cardboard."

I had this feeling on a regular basis once I began investigating the world's oldest people. After realizing that the Kimura case was nonsense, I began kicking the Styrofoam cactuses over with gusto.

Some cases required no effort at all. The former oldest woman in the United States, for example, was a former smoker and occasional drinker named Adele Dunlap. Adele just told everyone she was an error. In a farcical, tragicomically weird end to her life, nobody believed her, and she was collectively rubbished for being unable to remember her real age. Despite consistently claiming for almost a decade that her paperwork was wrong, and that she was a decade younger than her documents suggested, Adele was duly informed by newspapers that she was mistaken about herself.

The academics knew better. A typical exchange in *USA Today* ran thus: "asked how it felt to be 113, Dunlap . . . looked her questioner in the eye and answered: 'I'm 104.'" The reporter shot her down: "She's not. The Los Angeles-based Gerontology Research Group, which tracks supercentenarians, or people 110 and older, has validated her age."[31] Faced with a contradiction, Robert Young—director of the GRG that certifies the Guinness World Records and who claims he can see the thickness of black people's skin—told this woman on her deathbed that she was wrong and that she was certified as the oldest woman in the country.[32]

The logic was bizarre. It also made a lie of the idea that errors in age records will be discarded if, under cross-examination, any flaws are found in the biography of extremely old people. This woman, a black woman, spent her terminal years repeating on a loop that she was not the oldest woman in America. Nobody listened. The Obamas even sent her a birthday card to congratulate her on being 113. Adele Dunlap was a supercentenarian whether she liked it or not. She still is.

Other anomalies were kicked up by routine cases, including for people universally lauded for their age. It is hard to think of a more likable case than Richard Arvin Overton—the oldest man in the United States. Overton chain-smoked twelve to eighteen cigars a day, drank heavily, slurped Dr Pepper and coffee with relentless gusto, treated fried catfish and ice cream as staple foods, and both started and ended his day on the whiskey.[33] Yet compared to other cases, his evidence seems relatively clear and well vetted, again by Robert Young, so it is worth revealing a case with the highest standards of evidence available.

Overton reports being born in Bastrop County, Texas. This is fortunate given that Texas has published its entire registry of birth, death, baptism, and marriage certificates back to 1903, but unfortunate as Overton isn't in the database. Neither are his family members or any other Overtons born in Bastrop County. Nowhere in the state of Texas, or at least in the central birth records, is he recorded, and so the oldest man in the United States does not seem to have a birth certificate.

This is almost certainly because, to quote the Texas State Library, "it was as late as the 1930s before births were consistently recorded."[34] This was typical of the racial inequality of America where (in the view of the racist) African American people simply didn't count and were thus not counted. It was also exacerbated by a general widespread absence of records nationwide.[35]

However, Overton does have some self-report census records to back up his age. They are, let's say, slightly funky. In the 1910 census, Overton is reported as aged four—consistent with his current date of birth May 11, 1906.[36] But ten years later in the 1920 census, he has aged only seven years, to become eleven years old—estimated as born in 1909.[37] Another ten years later in the 1930 census, he became twenty-two with an estimated birth year of 1907 and keeps this new age until he's thirty-three in the 1940 census.[38] Then his October 1940 US Army draft card moves him back in time again to be born in 1906, but he changes his birthday from May 11 to May 6 and weirdly reports that he has no middle name.[39] By the 1950 census, Overton is back to being reported as born in 1908, before he is reported again at his final birthdate of May 11, 1906. When asked about his age, Overton had cryptically said, "I didn't know when I came here, and I don't know when I'm going," which I suppose is consistent with having so many self-reported ages.[40]

Like Kimura-Miyake, this muddle contains at least three and more likely four reported birth years and seemingly no birth certificate. Yet this is about the best case I found. It is convincing to me that Overton fought with bravery and distinction through WWII and that the lovable bloke earned every second that he was feted in the White House by the Obamas. What is not clear to me is what age he was when he served or why his multiple self-reported birthdays and apparent (though not certain) lack of a birth certificate seem to never have been reported.

While such successful validated extreme-age cases were informative, so were the many incredible failures. Irma Borgoglio was one. She lived to 103 in the speck-sized and pleasant Italian village of Bellavista di Trivero with her unemployed son, 63-year-old Pierangelo Bussolera. The pair had been quiet enough, though Borgoglio was a bit of a shut-in, almost certainly as a result of her being over 100 and living at the top of several flights of stairs. The neighbors often saw her waving from her balcony at the top of the apartments. Some visitors had popped by, but Irma had been in bed.[41] After this happened a few times, and because everyone loved Irma, the local police decided to pop by and check on her.

When they arrived, Irma Borgoglio was upright in her kitchen. Pierangelo was claiming she was unwell, but the 103-year-old was as solid as ever. Indeed, she was even more solid than ever. She was frozen.[42]

Pierangelo had extracted his mother from the freezer and managed to get her "propped up in the kitchen" but had stumbled when he apparently tried to explain that she had just died minutes earlier. It hardly took Inspector Holmes to deduce why she was frozen next to an empty freezer.

Pierangelo had been unemployed and under financial pressure, and when his mother died, he seemingly panicked. Some arguing about money had been overheard by the neighbors, but it had "suddenly stopped in March."[43] The over-sixty Pierangelo had then told people, "She's not well, she needs to stay in bed for now," which was sort of true, and then bought a chest freezer and dragged it to the top floor apartment by himself. Later, visitors had noticed how strange it seemed to have a massive freezer for just two people, with so much stuff piled on the lid.[44] Pierangelo collected his mother's pension the whole time.

This kind of pension fraud is a common theme in Italy. A simple Google search in English, at just a single Italian newspaper, returns bodies: two more people in a freezer, one buried in a Sicilian garden by a caregiver, one found mummified in a wall, and one missing for eight years, all while their pensions were collected. This newspaper even has another false centenarian: a Roman woman who died between the unremarkable ages of 89 and 91 but who survived on paper to reach 104 while her son cashed the pension checks.[45] One reporter noted, drily, that "dead pensioners being kept so that crisis-hit relatives can continue collecting their benefits are becoming fairly frequent in Italy."[46]

These are just a few of the cases that made the news and were republished into English. Similar cases happen all the time, from China to Portsmouth.[47] In 2024 a pair of Irish lads were even jailed for carrying their emphatically dead uncle into the post office, like a marionette, to withdraw the princely sum of €246.[48] Following up this year was an Irish woman who was prosecuted for covering up her father-in-law's death until he reached 110, even putting her husband Séamus in his bed to play pretend when the Gardaí came over.[49] Half the town knew about it—they had the funeral after all—and had said nothing for twenty-eight years.[50]

The more boring variety of such frauds, in which people simply delay the reporting of a death, are far more common and usually don't get reported. The victims and details are hidden by the court, or the details are not quite juicy enough to make the papers.

Even extreme-age cases that satisfied the "gold standards" of validation, and were initially confirmed, failed without making the papers. For example, the first-ever centenarians to have their age validated using current scientific methods were discovered by the Taché investigation. Canadian civil servant Joseph-Charles Taché had officially investigated 421 claimants for extreme longevity, keeping 82 people with ages that were "proved by authentic documents, examined with a rigorous scrutiny."[51] As a little aside, the method of cross-checking documents is the only one available: there is no newer alternative.

The Taché investigation was in 1878. Of the 82 initially passed by validation techniques, how many have remained "proven" cases now? Well, only nine survived an immediate follow-up investigation. One was the world's (former) oldest man Pierre Joubert, one of the very many "first" people in history to survive past age 110.[52] Joubert was inevitably feted as the doyen of humanity and the certified oldest man ever at 113 years, until a century later when someone discovered he had actually died at age 65. You may think, of course, that the evidence must have been very obscure to lay so well hidden for a hundred years. No. It was written on his death certificate the entire time, and nobody had bothered to read it.

Of the nine double-validated centenarians remaining, Joubert and six others were found to be frauds or filing-cabinet phantoms. The seventh case was "left open in the absence of information allowing to eliminate it,"[53] and the last, of the initial 421 cases, is apparently still officially recognized.* Notable here is the apparent confidence that demographers still have in this last person. Sure, the other 420 people were fakes, but *this* time . . .

The validated and then debunked cases were often so obvious that they resembled folksy, larger-than-large Christmas movies with cheesy plots. In 1979 the oldest person in the Guinness World Records was the validated 137-year-old Charlie Smith, who had been "verified by Social Security officials who said they had turned up documents that showed Mr. Smith had been sold into slavery in New Orleans in 1859."[54] Opportunity had leaped on Charlie when he was picking fruit, and his employer told him he had to get some kind of ID. The Social Security Administration found him an ID on file, promptly informed him he was 113 years old, and then lost the file.[55] The old gas-bagger just went along with it and, hilariously, the entire nation bought the jibber-jabber.

People believed Charlie no matter how big the tale. He said he was the last surviving American slave, last Union soldier of the American Civil War, and the first African American to be made a Freemason.† He came from "Galina, Africa": a port in Liberia that doesn't exist, where he was apparently lured onto a slave boat by a "syrup tree" and a slave master named Legree.[56] He said he shot an innocent preacher in a packed church and the judge just let him go. Charlie also had a lot of ever-changing jobs, always more glamorous than fruit-picking, like lawman, bounty hunter, train robber,

---

* Anne-Charlotte Dumont, who died at age one hundred years and one month.

† This was Prince Hall and fourteen other anonymous African Americans in 1774, followed by literally thousands of other people, well before Charlie was claimed as entering the United States in 1854.

and honkytonk owner. He said he was a Texas cowboy who shot "bears and panthers . . . and lions," an unusual addition to the local wildlife, and claims he rode with Billy the Kid to successfully arrest the man who shot* President Garfield.[57] Charlie had other troubles with anachronisms too, like in 1975 when he said he was 144 but was supposed to be 132.

Nobody cared. Instead of criticism, Charlie was invited to the launch of Apollo 17 as a VIP, where he loudly informed Neil Armstrong, Dick Gordon, and everyone else in the shelter, "Ain't nobody goin' t' no moon, you, me, or anybody else."[58] He was certain that the moon launch stuff was (ironically) all fake. As the rockets took off, he crossed to a studio audience with CBS and told Walter Cronkite "they ain't goin' there . . . ain't nobody runnin' on the moon" before confidently saying he arrested President Garfield's assassin. After seeing Charlie's charisma, Ripley's Believe It or Not! put him on national television and the BBC shot a documentary, with an amazingly posh British presenter throwing upper-class shade at all his tall tales. Charlie's life was then dramatized by a PBS film that uncritically supported his claim to have ridden with Jesse James, been a circus sideshow, and chased Billy the Kid.†

Like Billy and Jesse James, they never took him alive, and Charlie Smith was memorialized in *The New York Times* as "perhaps" age 137.[59] The entire nation loved it. It was all pure bullshit.

When I had begun checking the possibility of errors among the world's oldest people, this level of cognitive dissonance and selective evidence was confronting. Something that rapidly reinforced this impression was the lifestyle advice. These were supposedly the longest-lived, and presumably healthiest, people in existence. Yet where were all the athletes? Hikers? Military clean-living advocates, monks, ascetics, vegetarians, and day-spa regulars? Aerobically svelte, annoyingly pious, scare-you-with-their-abs types?

---

*The man who shot Garfield was arrested on the spot in Washington, roughly thirty seconds after the deed, 2,400 km away from Billy the Kid in New Mexico. Given that the Pony Express mail service took about ten days coast-to-coast, the mail about Garfield's shooting would have arrived about the same day that Billy the Kid was assassinated, making it rather hard for him to ride out in service to the law. But who cares, history schmistory, am I right?

†How Charlie went from chasing Billy, to become friends looking for Garfield's assassin, was sadly never clarified.

Dead.

Instead, the longevity record holders were heavy-drinking, heavy-smoking people from the rough parts of town. In a summary by the pro-extreme-age researcher Stephen Coles, they had lifestyles dominated by "heavy smoking, heavy drinking, or both, failure to exercise on a regular basis, and no conscious effort to eat nutritiously."[60]

Of the five oldest-ever men, first-placed multibirthday Kimura-Miyake is covered. Second-placed Christian Mortensen smoked and chewed tobacco continuously from age 20 to age 117, third-placed Emilio del Toro smoked for 76 years straight, fourth-placed Mathew Beard was arrested for drunk driving at age 90, and the fifth-oldest man Walter Breuning smoked cigars until age 108.[61] The oldest women are hardly better, just less well-documented. Four of the ten oldest-ever women don't have anything much about health written down, at least three of the rest smoked daily past age 100, and two drank daily. Adele Dunlap, the woman who fought against her own birth certificate, "never went out jogging or anything" and "smoked . . . and ate whatever she wanted" without raising suspicion.[62] Adele's fellow Jerseyite, Mississippi-born 111-year-old Aggie Fenton, engaged in 70 years of heavy drinking. Her longevity secret was to "drink three Miller High Life's and a shot of Johnnie Walker Blue Label scotch every day."[63] The United Kingdom and Europe's former oldest man instead credited "cigarettes, whiskey and wild, wild women."[64] Jeanne Calment, the putative world's oldest woman, ate a kilo of chocolate a week while smoking and drinking every day, spending more than a century on the smokes.[65]

Serious mental gymnastics are employed to excuse these counterindicators of health. The Max Planck scientists recorded Jeanne Calment smoking both one and two Gauloises cigarettes a day before concluding that she "possibly did not inhale at all" for over a hundred years.[66] The same conclusion was reached for the world's then-oldest man, Christian Mortensen, who chuffed at everything: "mainly a pipe and later on cigars, but almost never cigarettes . . . he had also chewed tobacco . . . but never inhaled."[67] Richard Overton, the oldest man in the United States, also claimed, somewhat more plausibly given cigar etiquette, that he didn't inhale his twelve to eighteen cigars a day: "I smoke my cigars; blow the smoke out. I don't swallow it."[68] Everyone naturally believed him, but sadly nobody asked if he spat out the whiskey. It appears that, like all the other heavy-drinking 110-year-olds, he swallowed.

These were supposed to be the most carefully validated age measurements on earth. They were obviously wrong, and nobody noticed. Many other people should have reached the same conclusion, including a normally skeptical brace of investigative journalists. Yet this was obviously not the case. Instead, the scientists involved were engaging in what appears to be a stupendous amount of self-deception.

I was staring at some special kind of madness, and I could not fathom the shape of it. I should have been working as a statistician in plant science, but I was staring at conjecture, shadows, and anecdotes. I found some errors but mostly found doubts. Whenever I read a case or a scientific paper, my reality lurched sideways.

I felt like I was dining with Kubrick, and boy was he drunk.

I needed something more concrete.

This brings me, absurdly, to the oldest dog in the world.

4 Paperwork of the Living Dead

Many years after I began investigating extreme-age records, my friend, who is inordinately fond of the four-legged tail-beasts, sent me an article about Bobi, the world's oldest dog. I was attempting to convince her that Bobi was probably invented. "That dog's a woofy fraud!" was the gist of my argument, along with the observation that dogs can't be interrogated about their life story or be relied upon to have accurate papers. However, as my friend is a brilliant investigative journalist, she wanted evidence. I only had a cynical nature and the *lack* of evidence: Bobi was a village dog who was "never put on a lead or chained up outside, instead being free to wander the village's streets," whose paperwork appeared legitimate but who could obviously have been swapped for any other dog at any time, or been mistaken for any other dog at any time, in his thirty-one years of claimed existence.[1]

The dog's owner, however, was adamant that Bobi was thirty-one—he had pictures of Bobi and paperwork registering him with the Portuguese authorities to prove his case—and the world got on board with the "wonder dog" story. Included in these true believers were, seemingly, all my acquaintances. The result was a large ensemble of relatives and friends who, once they caught wind of my doubts, gave me a good-natured teasing about my apparent crusade against the innocuous and quite sweet-looking dog. At some stage in this teasing I did, momentarily, reflect: what if I was the worst of all possible things, a dog-despiser? Of course I had doubts, of course I was skeptical, but was I overdoing it? Was my view, tinted by human errors, blinding me to the marvels of a longevous Portuguese dog?

I languished in doubt for weeks, as the story quietly died the natural death of a human interest story in the news cycle. The teasing tailed off. However, Bobi did eventually return.

The first little doubt came from social media. A photo of Bobi had been posted to his owner's Facebook in 2009 and then reposted by the Guinness World Records to support Bobi's life story. The only problem was that in 2009, Bobi had four white "socks" of fur on his paws, with the front-left sock going far above the knee, and white muzzle markings. Yet by 2023, Bobi had somehow magically "re-browned" himself: he had no socks, with chocolate brown legs and paws instead, and had somehow kept only a tiny fringe of white hairs on his muzzle in a new pattern and in the wrong place.

In a striking oversight, or a blind and credulous mistake, the offending fur-change pictures had been posted by the Guinness World Records website in support of Bobi's case.[2] Almost nobody had noticed that they were images of two very obviously different dogs. The internet rumors exploded, prompting an investigation by Matt Reynolds, a "journalist with an extremely high tolerance for low-stakes stories" at *Wired* magazine.[3] Reynolds dredged up Bobi's paperwork and discovered that the supposedly bulletproof case was more like a fragrant Swiss cheese. His government paperwork did not stretch back to his birth—it stretched back to 2022: Bobi had been registered with the authorities less than twelve months before he became famous. Guinness reportedly never even called to check Bobi's age with the Portuguese authorities that kept the dog database.[4] They just believed the bloke who said his dog was thirty-one.

All of this is even worse given that, to paraphrase the impeccable logic of dog-Reddit, even if you have legitimate paperwork for one dog, you can just "swap him out" for a younger or not-dead dog to make it appear to live longer. Essentially, if you find a similar-looking dog, pull out your old paperwork, and proclaim "that's my Bobi," there is no way to refute you. Bobi could have died twenty years ago—it's not like dog death verification is high on the government agenda—and you can put a fresh pooch in his place every year if you fancy. Nobody will notice. How else do you suppose that the television dog show *Lassie* was consistently on-screen from 1938 to 2006? Lassie is nine sequentially-deceased dogs.

The true believers were not so easily shaken off, however, and the scandal dragged on for months. The response from Bobi's owner settled on blaming a cabal of "the 'elite' whose objective is clearly to discredit Bobi's life" so that they could sell pet food instead of letting dogs eat human food like Bobi.[5] The secret conspirators that run the world never cease to surprise me: apparently dyeing the paws of a geriatric dog to trick a Portuguese man was now on their to-do list. I wonder if Big Barry's Beef Biscuits is pulling the strings of global power?

To be fair to the proposed cult of some eighteen thousand veterinarians devoted to caring for animals, their oddly non-Satanic-looking spokesman Dr. Danny Chambers of the Royal College of Veterinary Surgeons had raised these doubts when the case first broke. Confronted with the possibility of Portuguese Super-Pooch, Danny had stated flatly that "not a single one of my veterinary colleagues believe he was actually 31 years old" and suggested the case was "completely implausible."[6] Danny, eighteen thousand vets, and I—we seemed to be the only doggy-dog cynics.

Danny was roundly hated on social media at the time, in keeping with tradition, but he was ultimately vindicated. After a delay, Bobi was finally ruled an impostor. The Guinness World Records revoked Bobi's claim and reverted to an equally tenuous, but less testable, claim for world's oldest dog. Seemingly spooked by the affair, the Guinness staff have also been confronted by a central problem in longevity research and "are reviewing how we verify animal age records at the moment."[7] The only method of pet age validation involves paperwork, and as every government in the world knows and cannot remedy, paperwork can be easily swapped or stolen or forged. Good luck to them, I suppose.

The challenges of validating Bobi the dog seem frivolous, and they are, but they betray the way in which we measure biological ages in many long-lived species. For organisms like trees and mollusks and a few other cases like *Tyrannosaurus rex*, we can count growth rings to measure age.[8] In general, however, when we cannot continuously observe a species for its entire lifespan, we can only use paperwork and records to track the age of species. But paperwork is easily mistaken or forged or lost or stolen; it is prone to mistakes and errors. What validation looks like in demographic practice is therefore not validation in the measurement theory sense, in which you check something against another independent and accurate metric or against a fixed standard like the standard kilogram. Validation in demography is merely checking to make sure a set of documents and records are all consistent with each other.[9] The documents must record the same age, consistently, and that's good enough.

Even this very limited test is routinely altered or ignored. In the case of Kimura, multiple birthdays were reported, but each was waved away because nobody felt like they mattered. The same thing happened with Jeanne Calment: she has different ages recorded in her census returns, but despite that this technically violates the practice of "validation," the validating demographers simply choose to ignore the discrepancy based on their own judgment. The same process is also likely to happen with the five or so reported ages of Richard Overton. Discrepancies don't matter if you *feel like* they don't matter, and that's it.

Even if this was not the case, if your documents are consistent but wrong, well, there is no protection against that either.

Back in 2018, this idea prompted me to make a thought experiment to explain what I had been seeing in the extreme-age records.

Imagine you had a room. In that room are one hundred real, bona fide centenarian women with all their government birth registration paperwork in hand. All the paperwork is real, valid, and up-to-date. Now imagine you swap just one of these people with their younger sister and hand that sister the (real and validated) paperwork.

The challenge is how would you detect this kind of error?

You can't do it by looking at the paperwork because birth certificates don't have photo ID or fingerprints, even today. The paperwork is all real, government issued, and perfectly consistent. You also can't do it by interviewing the person: they will know enough about (say) their older sister's life to pass interview or, if they forget something, plausibly claim a memory lapse. You cannot really discriminate between a forgetful or dementia-suffering 105-year-old and the uncertain memory of a 95-year-old sibling.

So, how do you detect this swap in a way that is reproducible and accurate?

The answer is that you cannot.

Now imagine that I have two rooms of claimed centenarians. One is filled with one hundred real centenarians with their paperwork and the other with one hundred "sibling swap" fakes, holding real documents. How would you discriminate, in a reproducible way, which room is which?

I asked these questions to dozens of scientists and even published the conundrum.[10] So far, nobody can tell me an answer. The conclusion must be that even with a 100 percent error rate, certain errors are undetectable to existing methods.*

* Although some demographers have pooh-poohed the idea as fanciful, sibling swaps are pretty routine. The former "world's oldest men" record holders include people who were their own younger siblings, like Jesus Arenas (age 118) and Shigechiyo Izumi (age 120), and some who were recorded as their own fathers, like Pierre Joubert (age 111). None of them were detected during validation or interview.

Adding to this undetectable error-generating process is the Carrie White style of error, where an incorrect birthdate is written down and copied across multiple documents, which of course all become perfectly consistent with one another. The victims of such events don't require a younger sibling to make an error: they often don't even know a mistake has happened to them because, even in "developed" countries, the literacy rate one hundred years ago was often single digits. For example, in 1900 when the Italian centenarians went to school, Italy had an average of 2.8 years of schooling per person, the population was at best about 50 percent literate, and the literacy test was to sign your own name.[11]

As we have already seen, these potential issues are further compounded because almost nobody performs any interviews and even when they do, they frequently just ignore any contradictions (like having three birthdays or claiming you are ten years younger) so they can publish the data.[12] After all, there are often career rewards only if you overlook the errors: it is pretty hard to publish a scientific case report about a grandma reaching 88 years old, but if she manages 122 you might even squeeze a book out of it.

With this thought experiment I had finally grabbed onto something concrete. I had an admission that errors were undetectable, I knew this kind of substitution occurred, and I knew that there was no ceiling to the potential error rates. Yet beyond that, I had very little. I had to keep digging.

One more case, however, was enough to blow my suspicions into the stratosphere.

"Respect for the Aged Day 2010" was coming up, and officials in the megacity of Tokyo were tallying up a list of the local elderly. At the top was Sogen Kato, a 111-year-old gentleman. The city officials, naturally, decided they wanted to celebrate him. However, whenever city officials went by his house, Kato's family had strangely claimed he was out for the day, in Gifu Prefecture, five hours' travel away. Repeated house calls were made, with various members of Kato's family answering the door to tell the officials that Kato "doesn't want to see anyone" or that he had just gone out.[13]

To Gifu. At age 111.

Eventually, one of Kato's grandchildren turned up to a police station with a strange story. The 111-year-old had apparently barricaded himself in his room. She revealed that "Grandpa was a very scary man. So, we couldn't open the door. . . . He shut himself up in the room without food or water."[14]

It seemed a remarkable physical feat for a 111-year-old to fight off his collective family, so the police came by for a look.

Kato was indeed barricaded in the room. He had been in there for at least thirty years. He was also dead. The other relative in the backpack at the foot of the bed was likewise dead, but it was not clear for how long. Kato was a mummified husk, though his family steadfastly claimed that he was in a meditative transcendental state. He had supposedly wanted to become a living Buddha and had reached "sokushinbutsu"—a state of self-mummification. No mention of the bag-resident's religious status was put forward, unfortunately.

The entire extended family seemed to be in on it. Four of them even lived in the house with the corpses, while, in sharp contrast to his body, Kato's pension had stayed very much alive and had been collecting itself with striking regularity.[15] At least 9.5 million yen* had been paid out in the past few years, and Kato's mummy had even withdrawn 2.7 million yen in a lump sum after the city officials came calling.[16]

This was all pretty routine for a supercentenarian case, but what happened next was not. First, officials checked on the oldest woman in Tokyo, the 113-year-old Fusa Furuya, who turned out to have been "missing" since 1986.[17] This prompted the Japanese Ministry of Justice to check in on other centenarians: not just in Tokyo but across all Japan.

When the Ministry of Justice finished investigating, 234,354 supposedly "living" people had become dead.[18] Over 82 percent of all the people in Japan aged one hundred and older were discovered to be dead like Kato or "missing" like Furuya. This is even more remarkable given that the authorities did not manage to check everyone, with about 10 percent of family register records unchecked, partly because only "about 80 percent" of municipal records were computerized at the time.[19]

This led to what I call the anti-Calment law where, in a fit of legislative weirdness, it became possible to ask the local mayor to "remove" (legally kill) elderly household members, on the condition that "the person must be an elderly person of 120 years or older" and that they cannot be located.[20] The initial investigation had killed 77,118 people aged over 120 and I often wonder how many more paperwork ghosts have been caught and killed by the mayoral death squads. Unfortunately, I haven't been able to find out.

---

*About $110,000 in 2024 dollars.

The response of the scientific community to this was muted, if not monkishly silent. In 1994, Japanese extreme-age data from the Ministry of Justice was ranked as equal best in the world.[21] In 2008 these data had been triumphantly hailed as "among the best in the world" by leading demographers, in a paper titled "They Really Are That Old."[22] In 2010 these world leaders were missing or dead.

Imagine if 82 percent of the stars seen by the Hubble Space Telescope turned out to be dust on the lens. In astronomy, or any other science, this would have caused a massive scandal. In aging science there was not even a shrug. The scientific community basically murmured "sure, but the *average* star brightness barely changes, so who cares" and moved on.

It seems only one paper examined the crisis in the aftermath.[23] It concluded that the disappearance of so many records was immaterial because "the impact of this group on life expectancy statistics, however, is effectively nil."[24] In other words, you could ignore the scale of the problem if you simply hid the errors by averaging them across the entire population of Japan. Group averages were never the issue. The issue was that people had been scientifically examining the remarkable rate of centenarians occurring in Japan for decades—while proclaiming the high quality of the data generated by the Ministry of Justice documents—and the people were dead the entire time.[25]

My fun but decidedly weird investigation into the extreme oldest old had expanded. To my surprise, I had discovered something of actual relevance. This was not just a collection of anecdotes. It was a systematic disaster.

To understand the scale of this problem, it is necessary to explain how old-age mortality data and models are used to allocate trillions of dollars in global investment.

Pension rates—the amount you are asked to or forced to save for retirement—must be determined when you begin your working life when you are young. Governments, individuals, and businesses therefore need to estimate how much money to ask (or force) people to save and whether paying 9.5, 10, or 11 percent of your wages into a retirement fund will be enough to allow you to live from retirement to death seventy years into the future. The estimate for how much an average person needs to retire therefore relies on a distant projection of how long people will live.[26]

To do that, global governments and pension funds assume that survival rates will continue getting better and that extreme late-life mortality *now* will be representative of future survival patterns *later*. We look to the extremes of survival today to model the expected survival of future societies, in the encouraging expectation that the extreme outliers of longevity today will be a more typical survival rate in the future.

Extreme-age data and models today are therefore our primary basis for understanding and predicting future populations.[27] The United Nations (UN) even provides these estimates as a core service, from now until one hundred years into the future, for every government on earth.[28]

There are other important uses for such data. Governments also need accurate estimates of late-life survival to understand how many eighty-, ninety-, and one-hundred-year-olds they will have to take care of in the decades to come, and they spend huge volumes of our money based on these projections. Large-scale aged care, hospitals, and infrastructure all need to be planned a decade or two in advance, and models built on extreme-age data are used to project that demand into the future. It takes ten years or more to build a hospital, from planning to opening day. Planning a health system needs even more foresight. This is a central purpose of demographic science and good government. It also depends on having accurate estimates of old-age survival.

Another way in which these data touch our daily lives is through insurance. Everyone who pays a life insurance premium is being allocated a risk of death by a number cruncher. Insurance companies and pension funds are obsessed with "longevity risk": the (ironic) risk that their customers will survive too long and will therefore pay more into their life insurance, or need a larger pension, before death. As your death is usually decades away, such organizations extensively model the odds that humans will survive to much older ages in the future. Longevity risk inspires both fear and intensive research in economics and insurance, and is a major determinant of how much people across the globe pay in life insurance premiums and pensions day-to-day.[29] This adds up to trillions of investment dollars over time, and these rates are largely fixed by governments and firms in London who listen to my colleagues at Oxford.

To get these data systematically wrong is therefore a *cataclysmic* blunder.

If the damage had been contained to Japan, it would be disturbing. Yet the Japanese case was not a one-off. I wished it were. Instead, I immediately discovered another population of dead people.

The Greek government had gotten itself into trouble during the global financial crisis. As a result of the disaster, and its need to raise its credibility and credit rating, the government began tackling systemic problems with its notoriously dicey public finances. In the process, it discovered that around two hundred thousand people (one in every fifty) had been committing pension fraud, including people claiming benefits for blindness but driving taxis and renewing their truck licenses at the same time.[30]

The state had also been paying over nine thousand pensions to 100-plus-year-old Greeks. This was not a problem, in and of itself, until the ministry tried to show that they were alive. Curiously, the Greeks had already investigated centenarian cases in 2010 and found only 321 errors.[31] It was only after the German minister responsible for helping bail out the Greek government threatened "the need to examine every last inch of Greece's finances, before even one cent (in aid) is provided" did they discover another few thousand cases.[32] The centenarians evaporated overnight. According to the Greek minister responsible for paying the pensions—who has probably the greatest incentive of anyone to minimize the scale of problem—virtually every centenarian in the country was a product of pension fraud.[33]

Like Sogen Kato, the Greek centenarians were only alive on pension day. I could not get an exact number, but at least 72 percent of them disappeared after the 2011 census. You can, of course, lie on the census returns and just say you have a 100-year-old in the house, especially if you want to continue collecting a dead person's pension, so take this as a minimum estimate.

This time there was no response by the academic community because, despite being published in national newspapers, nobody in academia even noticed. The studies of the oldest old just carried on regardless, and years after the crisis, I got to be the first academic to publish the mix-up.[34] I would find more populations, like the thirty thousand Italian citizens who collected their pension from the grave, but the damage was done.[35] I knew the records for the world's oldest people were flawed and that massive error rates could lay undetected, for decades, in supposedly world-leading data.

Troubling questions slowly crept out at me. How big was this problem? At what age did our medical and age data stop being reliable? What other countries were affected, and how deep did this problem go?

I began to track down the world's oldest people one by one.

5 Chattahoochee Tenants

My first step was to find a comprehensive sample of the oldest people. I gained access to the two largest databases of extremely old people in the world, and the start was not encouraging.

The first database was the Gerontology Research Group (GRG) out of California, which I had first encountered when I debunked the *Nature* paper that rounded everything off to zero. The first problem in the GRG data was an immediate lack of any real documentation, scientific processes, or statistical plan. The GRG was essentially a loosely curated collection of Excel spreadsheets on an early 2000s wiki page. There was no version control, no registration of the protocols, and no record of what had been edited or why.

There were also problems with how the records were discovered. The general approach to finding cases was to have volunteers scan newspapers, looking for people announced as reaching a remarkable age. This introduces what is called ascertainment bias: you only find cases where you look. For example, at the time the GRG listed no Korean volunteers, and Korea therefore had none of the world's 110-plus-year-olds. Japan was four times the size, and has about the same average life expectancy, but had well over 120 people aged over 110—almost certainly because they had Japanese-speaking volunteers. However, I was going to eventually find out that there were much worse biases lurking in the undergrowth.

The database was maintained by Robert Young, who was promoted to the director of the GRG database after more than sixteen years spent validating the world's oldest people. He has validated about half of the world's 110-plus-year-olds, is a contributor to every major database on extreme longevity, and was key in establishing the New England Centenarian Study—arguably the most important centenarian study in the world. Young's capacity for fair judgment—the thing that case validations rest on—should therefore be unimpeachable.

Instead, he holds some irregular opinions. His master's thesis alone is enlightening. Some problems are simple, like the observation that "people who look young for their age live longer (citation needed)" [sic].[1] Others are more problematic.

Young promotes resurrecting race science using what he calls the "biological superiority hypothesis"[2]—which he supports despite it being not "politically correct"[3]—because he "gained a better understanding of the myth of African-American longevity, mainly from the movies and P.T. Barnum" and quotes Eddy Murphy's comedy *Coming to America* as a source material.[4] Young then compares living in Africa today to being Jewish in Auschwitz—saying that the genocide of the Jewish people was "akin to cutting down a tree, not pruning it"—before he then suggests he can sense the thickness of black people's skin.[5] Young goes on to surmise that African Americans may survive to remarkable ages because their "thick skin" protects their "internal organs." His primary evidence is that there is "a saying in the black community, 'black don't crack.'"[6] Young even does some Olympics-watching and finds that in the 2008 sprinting events, "blacks took not just first, second, and third—no non-Africans can be seen on the track. . . . But so far, no research has demonstrated that the ability to run fast is greater for one race of people than another due to genetics. This is something that, perhaps due to political correctness, has not been studied."[7] Ignoring, of course, the studies of the Third Reich, which clung to similar reasoning when Jesse Owens destroyed them in athletics. Young then concludes that Africans may run faster because they had "to escape from wild animals but also other humans," which also mixes in his constant confusion between physical fitness and Darwinian fitness.[8]

This would all just constitute a lousy thesis that should have never passed examination—I am looking at you, Georgia State University—if it were not for Young's ongoing importance in longevity research. Young has validated roughly half of all extreme-age data, but his reach extends further. For example, he was also responsible for assembling the New England Centenarian Study, the "gold standard" clinical study on centenarians in the United States.[9] To establish the study, Young decided who was and was not one hundred years old. He then screened cases and played "key roles in subject recruitment, ascertainment of demographic data, and age validation."[10] His method was to gather cases by scrolling the internet for news stories and waiting for emails—unsolicited emails—about old people.[11] Despite having no way of establishing the level of bias introduced, the study authors concluded that "ascertainment bias is likely minimal."[12] All of which boded poorly for the role of hidden biases in the GRG, where Young's "preferences" might creep into the data he was assembling.

In contrast, the other database of very old people had some hallmarks of respectability. This was the International Database for Longevity (IDL), established by Professor Jim Vaupel, who fought one-half of the debate in chapter 2.[13] I had my differences of opinion with Professor Vaupel himself,* including when I debunked his *Science* paper in chapter 1, but I generally held a high opinion of his staff and their data.

The IDL database was more coherent and seemed more careful.[14] There was some semblance of accountability, and the methods were published. Of course, there were still oddities and problems: for example, days of birth were hidden for certain countries without any explanation. The database also inevitably included many of Young's validations and listed him as a major contributor. Yet in general it seemed the more systematic of the two databases, and I got down to work filling in gaps and tracking down the records.

Before long I had traced the locations of birth and death for some 80 percent of the world's oldest people, focusing on anyone who lived in countries with enough records to perform a meaningful analysis. The cleanest way to compare public health is within a country, as its regions have a shared history, health system, and recordkeeping legislation. Many of the oldest people lived in countries with too few records to make comparisons, but most lived in places that (obviously) had good records and many internal regions: Japan, Italy, France, England, and the United States.

Long before I finished compiling the records, I began to feel I had been hoodwinked by prestige. The gold-standard IDL database had thousands of records but only "exhaustively validated" a few hundred people. Even among these gilded few, most did not have a birth or death certificate,

---

*In my opinion, and by his coauthor J. M. Aburto's seemingly clear and open public admissions, Vaupel and Aburto open-reviewed my ideas one month and published the same data and idea as their own the next, with no citation or acknowledgement of my work and the usual lack of consequences. Leaving me to ask: How would one more case, in a long line of banal and openly perpetuated cases of plagiarism I have witnessed in the social sciences, look any different to this?

which is weird because 90 percent of these people had died since 1990, in rich countries where 95–99 percent of citizens received a death certificate. The French and English records were all born into countries where a birth certificate was a legal requirement for over 120 years, yet they didn't have a birth certificate. For example, some 9,878 French people were recorded as age 105-plus in the IDL. None of these 9,878 people were marked down as having an original birth certificate,[15] and only 365 (3.6 percent) were "exhaustively" validated.[16]

Why? Well, nobody had ever wondered why, which is strange given that this database had been curated for decades.

It got much stranger.

I quickly mapped the location of the very old people against basic indicators of health, like gross domestic product (GDP), poverty, and crime rates: stuff that an undergraduate might study in "population health 101" or that the *Economist* or *Financial Times* might craft into a nice infographic for their average readers. The world's oldest people routinely came from the worst regions by each of these indicators.

In the United Kingdom, more people aged 105 and older came from Tower Hamlets—a poor inner-city suburb filled with tenements—than all the rich and posh regions of England combined. Tower Hamlets had "the highest poverty rate, highest child poverty rate, highest income inequality, the shortest disability-free life expectancy by ten to fifteen years, and the worst index of multiple deprivation" in London, and yet somehow it produced the highest density of extremely old people in the country.[17] Tower Hamlets was also the single worst place in all of England for old-age poverty, had the highest crime rate, and strikingly had the fewest 90-plus-year-olds out of 317 English districts. This last number was very strange given that the count of 90-plus-year-olds also includes all the 105-plus-year-olds. Despite this, Tower Hamlets stayed top of the charts.[18]

Similar oddities played out in France, where most extremely old people came from overseas territories and colonial holdings that were routinely left behind by the French government—regions with terrible rates of employment (50 percent in some regions), the highest rates of poverty, and crap recordkeeping by a historically indifferent colonial government.[19] These issues were extra odd given that huge numbers of people had migrated out of these territories in the 1960s. The people who stayed behind, in poverty, made it to 110 at stupendous rates. Not a single one of the out-migrants to the much richer and more carefully recorded regions of mainland France survived past 110.

To give a concrete example, over four hundred thousand internal migrants descended from Martinique and Guadeloupe live on mainland France: roughly equal to the population that still live on the islands. Despite migrants consistently being much healthier on average than nonmigrants, none of these people ended up reaching 110.[20] Guadeloupe and Martinique have eight 110-plus-year-olds each. They also have extraordinary rates of poverty compared to mainland France.

A cross-cutting finding then came from the United States, where the introduction of birth certificates happened bit by bit and state by state over time.[21] For example, the last states to get birth certificates only managed the feat in 1956. On an analytical whim, I took the rate at which each state produced 110-plus-year-olds and lined it up against the year the state legally introduced birth certificates. At the time most states introduced birth certificates, the United States was undergoing massive population growth and huge improvements in life expectancy, so you expected to see a massive rise in the number of people reaching 110-plus years old. Despite this, the introduction of birth certificates was linked to a roughly 80 percent reduction in the number of 110-plus-year-olds being born in the country.

The world's oldest people had birth certificate allergies.

The emerging picture was the same everywhere.[22] Regions in Japan, Italy, and the United States that were consistently rated as having the worst old-age health produced extremely old people at the highest rate. In short, there seemed to be a fundamental problem for every metric, country, and province.[23] Perhaps the peak absurdity was when I found that having *more* 90-plus-year-olds in a region predicted having *fewer* people who were older than 105 or 110. Within Italian cohorts (a cohort is the same group of people, followed over time), a higher chance of reaching age 55 predicted a *lower* chance of reaching ages 100-plus, 105-plus, and 110-plus.

Most damning in a cultural sense was the fact that the GRG data failed a basic demographic test for data quality. People who invent ages (e.g., when faking a birthdate or committing pension fraud) have a weak tendency to round off fake ages to certain dates: most often the first day of the month or some multiple of five or ten. The GRG, the extreme old database curated by Robert Young, had badly failed this test, indicating that these "exhaustively validated" data were filled with undetected fake records.[24] What was most disturbing about this was not that it happened but that it had not happened *before*. Age heaping is revealed to undergraduates in introductory demography courses as a basic tool for measuring data quality. Yet despite decades of published studies using these data, nobody had ever bothered to apply the "Demography 101" method to the database, and they reacted with shock or (most often) denial when the data failed.

To me, this all made a coherent story. While it was painful to dig into every individual case, there seemed widespread issues with even "exhaustively" validated people. When you expanded the analysis and collected the extremely old people together as a group, the patterns became worse. None of the extreme-age records made sense, and the best explanation was that these data were a collection of statistical junk.

Satisfied that I had given the data a thorough and comprehensive analysis—I had used the gold-standard databases, almost all the 110-plus-year-olds in the world, and reproducible code—I sent my findings off to a midtier respectable journal and attempted to get back to work as a plant scientist.

After about four weeks of quietly writing code about plant experiments, I remembered the paper and checked in with the editor.[25] While I was there, I did two little things. I put the paper up on a preprint server, a kind of nerd-wiki page where scientists post early results to say "I did it first," and then tweeted the preprint to the two hundred or so friends, family, and bots that followed my dull social media feed.[26] It got fifteen likes. It was the only self-promotion I would ever do.

My peaceful oat-yield code-writing lasted about another six weeks. The paper was desk rejected without review, which is common enough. Then, over a coffee and an Italian donut, I saw the first real sign that my slow-burn depression was going to get kicked out for a day or two. I got an excited email from a colleague saying I was featured on a statistics blog—scientists get excited about some weird things. I clicked through, saw some strange comments and enthusiastic debate, and decided to check the nerd-wiki with my preprinted paper. My funny little paper had over one hundred thousand downloads, and I was being seen by millions of people on social media. By the end of the week, it was tens of millions. Scientists outside the field were joking about the irony of my results, and the public joined in enthusiastically on social media. I had achieved the sad trappings of nerd fame: people cared, somewhat, about my research.

I began getting unprofessional hate mail immediately. Robert Young, the GRG director, was particularly forward. Young seemed intent on breaking Ayn Rand's world record for hiding the smallest amount of thought inside the largest amount of ink. Not wanting to let his efforts remain private, he began saying I was a "liar" and a "fraud" with "questionable motives," a self-promoter who was "greatly flawed, unfounded, uncorroborated, and unhelpful."[27] I especially enjoyed the ad hominem attacks that I somehow faked or never received my doctorate, a piece of imagination that does not seem to have gone away.[28] Other brilliant lines of reasoning suggested I was committing fraud by somehow using open code and other people's data to fool a rapidly growing audience.[29]

Yet Young kept demanding I send him my "evidence."[30] I sent the man the link to the code I had posted alongside the preprint and tried a few times to explain that he might consider reading the code and methods before making public allegations of fraud. Or just find someone else who could read code and look at the methods. After a few months of silence, Young came back even more angry. He had been unable to find the "correct app" to open the file I sent him: a plain text file, written in a common programming language. He then reached a conclusion that I must be lying because "the alleged supplementary file is only 27KB . . . there is no data* there" and followed this up with a demand that I instead send him my statistical analysis as an Excel spreadsheet.[31] This request made about as much sense as admitting to a librarian that you do not know how to open a book, cannot read Spanish, and therefore require your copy of *Don Quixote* in the form of a pie. Such a thing may be possible, I guess, but the request itself was deeply baffling. At this point, I decided it was best to just move on with my life.

Fortunately, not much of this had any personal impact. Professor Bob Furbank, my long-suffering boss in plant science, codiscoverer of the Furbank/Von Caemmerer model of C4 photosynthesis and a generally fantastic bloke, had sort of chuckled, shrugged, and asked me about wheat yields.[32] Bob knew I had recently blown the whistle on the Australian government for diverting tens of millions from a food program to the alcohol industry, and survived the litigation-based attempt at silencing and prosecution,[33] so

---

*My R code called out to a website and downloaded the required data.

he knew I was a bit of a cheeky scamp. My other boss, Professor Eric Stone, called me in to ask me with alarm why someone named Robert Young was emailing him and demanding I immediately be fired.

I calmly explained to him what had happened and optimistically insisted it would all blow over. It would not. Instead, the would-be graverobber Dr. Nikolay Zak was about to arrive.[34]

Dr. Nikolay Zak had published a simple hypothesis. He proposed that the world's oldest woman, the supposedly 122.5-year-old Jeanne Calment, was a fraud. He presented nineteen pieces of evidence, all circumstantial, to support an opinion that Calment's long-term unemployed daughter Yvonne had deliberately registered the wrong death and stolen her mother's identity to avoid the steep French inheritance taxes.[35] The upshot being that Yvonne had survived, pretended to be her own mother, and generated a fake age of 122.5.

Dr. Zak's paper went viral, but like my own recent experiences, the reaction to this proposal was uncomfortably based on personality over evidence. In the words of *The Guardian*, "Zak wasn't even a gerontologist, a specialist in aging, but a thirty-six-year-old mathematics graduate who worked as a glassblower at Moscow State University and hadn't published a paper in 10 years," which was true if you ignore his gerontologist coauthor but was rather immaterial.[36] For example, if a glassblower found a new packing of spheres in geometry or a new subtype of quasar in astronomy, the takeaway would be that it was *more* impressive than a professional scientist making the same discovery. Instead of calmly issuing a rebuttal, however, the finest minds in demography reached other conclusions.

A few scientists, backed by a former Central Intelligence Agency agent, suggested that Nikolay Zak and everyone linked to him were spies running a Russian disinformation campaign—with adjectives like "conceivable" used to achieve the required legalese arse-covering that protects journalists from defamation lawsuits—in *The Washington Post*.[37] The French newspaper *La Parisien* jumped on board, and hit the supposed conspirators with accusations of "fake news" backed by Russia.[38] *The Guardian* wondered out loud, "Was the Russian trying to sow doubt, so that his countrymen could take the lead in the gerontology field?"[39] To which the answer was obviously no, but the allegations stuck fast anyway. As Nikolay Zak had achieved viral fame through a preprint just after me, the obvious parallels allowed everyone to lump us into the same camp, and I was accused of being "in on it" too.

The person who had apparently unraveled this unlikely spy network was none other than Robert Young. How he achieved this amazing piece of spycraft is still largely a mystery, to me and probably to himself, but we can follow the loose outlines.

In between struggling to understand Excel spreadsheets, Young had begun to unravel the deepest plots of Vladimir Putin by spotting "unusual activity" on his self-managed wiki fan site, where "random accounts had been popping up on the 110 Club, an online forum dedicated to supercentenarians."[40] A Wikipedia page had also been anonymously edited, probably by Zak or his coauthors, to highlight the daughter-swap theory. While these may resemble everyday occurrences to us, Young instead recognized the workings of malign and evil forces, concluding there was "an intentional Anti-Jeanne Calment (anti-France=anti-EU=anti-West)[sic] disinformation and propaganda campaign coming from Russia and it is targeting many outlets."[41] By "many outlets," he seemed to mean Wikipedia and the wiki fan-page he administered.

Young was not alone in his genius. Professor S. Jay Olshansky also saw conspiracy, supporting a conclusion that "the Russians . . . provide misleading information on longevity" by citing debunked longevity cases from Georgia—which another demographer backed by raising similar cases from Azerbaijan.[42] Superb arguments, if not for two minor problems: the debunking was carried out by the Russian Professor Zhores Medvedev, and—in a secret known only to map-havers and book-readers—Georgia and Azerbaijan are not Russia.[43]

Such trifling details did not hold the spy catchers back, however, and neither did the broader problems. For example, some of the accused conspirators had been working in the United States as professors for decades: rather deep-cover agents to expose, just so the Federal Security Service of the Russian Federation could make Facebook comments about a pub trivia topic. Why would the Kremlin give a lumpy bottom-truffle about an academic dispute—and why would they use an easily identified kid and two lifelong Americans in Chicago to spread misinformation while blowing their (nonexistent) deep-cover identities—when they have a dozen untraceable troll factories on hand? Like the question "where is Azerbaijan?" these puzzles were overlooked.

The entire affair was a sad failure of journalism and reasoning. It also left me feeling like I was missing out on a nonstarting Lada sedan—bakłażan—filled with caviar and rubles. After all, if I were a spy too, where was my payoff? I'm still waiting, and I suppose Nikolay Zak is too, but I guess Putin is too preoccupied with genocide to pay up.

The only practical outcome of this controversy was to derail the entire debate into a fight about Jeanne Calment. I had posted my little paper after carefully building a case based on mathematics and theory and backed by evidence and the analysis of whole populations.[44] But instead of my original hope of discussing a widespread failure of technical demographic methods, or indeed the debunking of the world's oldest man, I was being dragged into a fight about old French photographs and an ongoing attempt to exhume a corpse.* Not a shred of reproducible evidence came forth. Once again, everything hung on opinions and dubious reasoning.

It made me feel like Bilbo the Hobbit, lost in Mirkwood. I wondered what I was doing here. Not for the last time.

I should have avoided the discussion about Calment altogether, but it was unfortunately too funny and erratic for me to just move on. I was also receiving weird threats in the mail. So now I must take a diversion from our core narrative to circle back and discuss the Curious Case of Jeanne Calment more thoroughly.

Jeanne Louise Calment was born in Arles, in what was then the Alpes-Maritimes region of France, where the average life expectancy was only 33 years. This was bad even for the time: a full 4.4 years below the national average. For perspective, the drop in Ukrainian life expectancy from 2019 to 2022, during the COVID-19 pandemic and the Russian invasion, was about 2.1 years. Calment reportedly died, also in Arles, at age 122.5 years old. She basically never moved from her house—living next door to a saucisson-slinging salami store called "Crazy About Sausages" that was older than her.† She never really worked, and neither did many of her family and

*I was only half-heartedly in favor of exhuming the bodies of Jeanne and Yvonne Calment because frankly I thought the whole affair to be unhinged and tangential, and because there was an untested vial of the lady's blood held at the Fondation Jean Dausset that could solve the mystery anyway.

†Sadly, "Foue de Saucissons" has been replaced by a real estate agency and, like Calment, is long gone into the nether realm. Nobody seems to have asked the big question, Did the sausages cause her longevity?

cohabitants, who lived off Calment's inherited money for decades. She is apparently relatively well-known, though evidence for this is taken long after the fact, and nobody records her 100th to 110th birthdays anywhere. It is also not at all clear why she was supposedly well-known given that she seems to have no achievement aside from her age.

Jeanne Calment was, like all the debunked cases before her, validated as the oldest person in the world through paperwork and interviews. Many people are rabid in Jeanne Calment's defense, some equally rabid in their doubts, but all parties seem to agree that her case is "broadly accepted as a golden standard of validation."[45] Her age is attested by multiple books, multiple documents, multiple investigating demographers, and many loud-and-proud French nationalists.[46] She even has a 2021 scientific paper with a title that ensures we know the author is a super-serious grown-up with big-boy pants: "The Real Facts Supporting Jeanne Calment as the World's Oldest Human."* The "real facts" consist of some slightly patchy but generally coherent witness testimony by Calment herself and "30 documents that the French researchers had found in the archives of Arles."[47] As a result, Calment's case is supported by a decent number of official documents: mostly census records that (approximately) track her increasing age.

This all sounds great. However, while I feel indifferent as to whether or not she is fake, I should detail the standard-issue set of doubts.

Her lifestyle was a huge issue and followed the supercentenarian standard of "generally terrible," none of which caused the validating demographers any concern. Apart from smoking for one hundred years and apparently not inhaling, Calment ate a kilogram of chocolate a week and drank every day, even in advanced age.[48] Other problems arise in the paperwork. Jeanne Calment's passport in the 1930s lists her hair and eyes as black.† Later in life they were listed, respectively, as somehow being brown and then light blue. Calment also has a few discrepancies early on in her census returns, like when she appears in two households at once. These are small issues, and while her paperwork is not as bulletproof as that of Carrie White (which was immaculate before she was debunked), Calment's papers generally seem to be in good order.

---

*I occasionally wonder why I don't add this disclaimer to my own scientific articles. Perhaps I could go further and state I'm "super-duper for real telling the truth you guys" to show the scientific community how cool and smart I am.

†There are no photos in most old passports, so descriptions are all you get.

Her *surviving* papers, I should say. Because as soon as she became famous for her advanced age, Calment grabbed her grandchild's cousin, who she had not seen for about two decades, and got her to incinerate all her personal effects and identifying documents. No realistic explanation has ever been offered for why Calment did this, and amazingly, the validating demographers do not seem to have been very interested in testing the very flimsy explanation.* Apparently, trying to burn a lifetime of personal effects at age 110 so that nobody can see them is, like smoking for one hundred years but never breathing in, a completely normal thing that regular people would totally do. The only vaguely coherent response to this question has come from an investigative reporter whose elegant counterpoint was "Why, if Calment were hiding something, would she have agreed to the validation interviews?"[49] A superb rhetorical question but not exactly a bulletproof defense. Criminals and senators will volunteer for routine questioning, after all, because bluffing looks less guilty than hiding. Just ask all the people in jail who voluntarily came down to the station "for a chat."

Added to these standard problems are an ever-increasing list of issues from Dr. Nikolay Zak and partners.[50] There are issues with these arguments, and they seem to become more problematic as they go along, so we can just skip ahead. This is all standard stuff, which has played out in the press for years.

Yet the biggest anomaly of the Jeanne Calment case is not the circumstantial evidence, like her lifestyle, or even the burning of her personal effects. It was something much more fun, odd, and unique. Something that everyone had missed.

Something hiding in the logic.

Normally, ideas in science are tested against some model of statistical likelihood. A hypothesis or idea is given more intellectual weight by the scientific community if it is considered as more statistically likely or more probable.

For me, this sets up the most enduringly delightful part of the Calment case. The part that everyone missed. Whether fake or real, Jeanne Calment's

---

*For Calment, the validating demographers were Professors Robine and Kannisto and the familiar face of Professor Jim Vaupel. They wrote books and launched careers off the case. This situation, where a scientist is directly benefiting from a case they are validating without any independent oversight, is what you might call problematic.

validated record is unique. Of the billions of humans alive, she is the only person validated to be 122.5 years old. Therefore, whether it is fraud, error, or some quirk of biology, Calment's age record *must* be caused by a statistically unlikely process.

The explanation for Jeanne Calment's age is therefore *allowed* to be implausibly weird and unlikely. So weird that if the situation were explained to any reasonable person, they would say "that would never happen" but with a tiny, tiny shadow of doubt: like if I told you a tactical-water-assault rabbit swam into the middle of a lake to threaten the US president, or that a crowd of Dutch people ate their disgraced prime minister à la carte, or that doctors are routinely breeding human-hamster hybrids as a fertility test.*

Short of the simply not possible, like aliens and lizard people, wild theories about Calment cannot be discarded on the basis that they are unlikely or improbable. In the well-reasoned and careful world of science, this qualifies as "fun."

This logical quirk eliminates all the objections to the hypothesis championed by Dr. Zak, that Jeanne Calment went through an identity swap with her daughter Yvonne. Jeanne Calment's entire household depended on Jeanne Calment, Yvonne had never worked a day in her lazy life, and Yvonne/Calment's death was certified by one Marie Nicolle: a woman *sans profession* in another country. So, let's have a faked death in a foreign sanatorium for tax fraud purposes. Why not? The idea is highly unlikely, sure, but it is allowed to be highly unlikely.

Permission to have an unlikely explanation also destroys many counterarguments. For example, Calment's demographers "found a fraud *improbable* [my emphasis] given the position and the life circumstances of the Calment family in this relatively small city of Arles where many people knew [Calment] and her daughter."[51] What happens to this defense—Calment was too well-known to commit identity fraud—if we are allowed an improbable (but not impossible) explanation?[52] Well, let's remember the tale about John Conway, who stole the identity of Stanley Kubrick in the middle of London. Calment is (shockingly) less famous than Stanley Kubrick: she was in fact an old woman with no immediate family, who

*Known as a humster, this man-rodent squib of cells is viable as an embryo for up to four days. I am desperately curious to know what the pro-life lobby, who think life begins at conception, has to say about humsters.

never worked and scarcely left the house. As theater managers and aspiring actors in the London gay scene would be able to tell you, she very well could have had her identity stolen, by her daughter or anyone else, and the public would not necessarily notice. Such an escapade fits our requirement of "highly improbable but not impossible" very well.

But why restrict our guesses to just one improbable idea and its counterarguments? Calment was alive for a long time after the death of her close relatives, but before she became famous (and burned her IDs and photo albums). Calment's last close relative, her grandson Frederic Billot, died in 1963, and she lived alone with no close relatives from 1963 to 1985. Her 100th birthday was recorded and reported nowhere. Neither was her 101st or 102nd birthday or any others before she was famous. She was sufficiently friendless that, at the time* there were no reports of her birthday all the way to 110. So, for 22 years, Calment was utterly alone in her home.

Anything could have happened. She could have fallen into the "foue de saucissons" sausage machine like a Sweeney Todd victim and had her ID stolen by a guilty sausage-frau. She could have had a lesbian love affair with someone who, seeing a rich, old, heirless lady with no close relatives and lots of money, carried out a masterful long-term identity switch. She could have had a never-do-well drunk swoop in, take her place,[53] and fool the entire town.†

All of this seems unlikely. So, it is exactly what we want.

That is what's so fun about Jeanne Calment. The case allows us to create slightly implausible explanations that sound vaguely made up.‡ While a mountain of ink has been spilled over Calment, arguing what is and is not

*There were some verbal reports made much later to explain away the absence of any contemporary records. They make very weird reading, explain virtually nothing, and raise a lot of secondary questions like "why was this lady so averse to people looking into her birthdays, until she was validated, at which point she talked to everyone?"

†In 1560 Martin Guerre won a court case defending himself against his uncle's accusation that he was an impostor. Most of the town, his four sisters, and his wife testified in court that he was Martin Guerre; he had, after all, spent years as the man, being father to Guerre's kid, and sleeping with Guerre's wife. Everyone was *certain* he was Martin Guerre. After Guerre won, the trial went to court again, and the original Martin Guerre limped into town on one leg: he had lost the other one at war. The two-legged, wife-shagging, sister-tricking Guerre was an impostor. People were gobsmacked.

‡A lot of supercentenarian cases seem to fit this criteria: many of them already have incredibly idiosyncratic backstories that border on the absurd.

likely, the underlying fun of the case has been lost completely. It doesn't have to be likely at all.

It is just one of the many pieces of logic that seem to have been missed.

Much of this just reinforces what we learned in earlier chapters: that a bulletproof-looking case ultimately depends not on some robust or reproducible test but on belief. The response to Dr. Zak's viral opinion piece was not a statistical or physical method but a collection of general feelings, and accusations of fraud and spying, coupled with the opinions that his arguments were, or were not, ridiculous. You could agree or disagree, but you couldn't reproducibly test anything.

You could only state your case, your opinion, about whether a hypothesis was *likely*, despite the obvious fact that you were looking at something *unlikely*, and pretend that this was somehow sufficient.

This brings us neatly back to our core narrative, after learning something of demographers but nothing certain about Jeanne Calment.

The motto of the Royal Society is *Nullius in verba*—it means, essentially, "Trust nobody's word" or "Take nobody's word for it." It is a scientific motto that embodies the spirit of both empiricism and mathematics: that you should only trust reproducible findings, not authority or opinion. With Jeanne Calment, I was being asked to take everybody's word for it, with no empirical test I could apply to discover the truth. I still could not put an old person in a centrifuge or a sausage mincer and measure whether they are 122.5, and I either trusted the paperwork or I didn't. Nothing had changed, and despite having just debunked the world's oldest man, nobody was questioning anything about the methods.

Hundreds of thousands of the world's oldest people were being discovered to be dead, and some of the "world's oldest people" records were just snake oil and wishes, and nobody was discussing what it meant. The world's very oldest people lived in places with high poverty and, ironically, few slightly less old people. Yet at no point was any explanation offered as to why these patterns arose or why almost the entire sample of the world's oldest people have no birth or death certificates. They were sprouting spy theories instead.

When a discussion of the population-wide patterns of data did eventually arrive, it was not the academic discourse I had hoped for. During my analysis, I had done a casual drive-by shooting of an idea called the Blue

Zones—regions that were supposedly filled with extremely old residents. The entire idea was flimsy, and debunking it was an afterthought. It had, however, prompted a surprisingly large number of questions by the public and the press.

I began to get emails. Lots of them. At first, they were just standard stuff: congratulations on making fun of something, unstructured anger without logic, and demands for retraction without any understanding of what I had done. Some people wrote to me to say what a nice holiday they had in Greece, while others wrote to tell me about their grandma. A few emails, however, came from a bloke named Dan Buettner who ran a lifestyle company. One day Dan took a turn and, in a case of emotional whiplash, went from very super-duper mad at me to super-duper best friends. He wrote to me, quite pleasantly, to invite me on an expedition.

It's worth quoting in full:

> "Hey Saul, I actually like your pluck and statistical prowess. Instead of focusing on a decade-plus old process, how about we team up to take advantage of more robust data and updated techniques to evolve the definition of blue zones? I've been commissioned by *National Geographic* to identify a sixth blue zone. We could discuss a new definition. I have a decent budget so we could do a good job—we could actually travel there and do the age verification together. Let me know. All best, Dan [sic]" (Buettner, pers. comm.)

I considered the lark, having never been invited on a *National Geographic* expedition, before I gave my answer. A definite no. You could argue that the offer resembles a bribe, and a rather meager one at that, so I was not very gentle.[54]

His overtures did, however, make me pay closer attention. Everyone was obsessed with the Blue Zones. So, who in the phonebook was this Dan Buettner character, and why was he offering me an unprompted romantic getaway?

6 Yeti-Hunting in the Blue Zones

The idea of a Blue Zone is deeply alluring. The principle is that there are regions of the world where people routinely live to extremely old ages, and if we want to live a long time, we should copy them. A sort of medical "when in Rome, do as the Romans do." Unfortunately, quantifying exactly what it is the Romans do and distilling it into lifestyle advice often involves reducing entire cultures to a bullet-point list: a kind of "nine culturally appropriated tips to make you live past 100."[1]

It is enlightening to trace ideas to their source. The first scientific paper mentioning a Blue Zone is the AKEA study, which identifies a few villages in the tranquil Mediterranean island of Sardinia as a region of "exceptional longevity."[2] Normally, discovering causal associations in medicine takes countless hours, hundreds of papers, and decades of careful experiments to unravel. Thankfully, the cause of the first Blue Zone was identified right away without such fuss: the secret to longevity was either "mountains" or "an alternative and interesting hypothesis . . . the high rate of inbreeding."[3]

That is, incest.

This "put a bun in the oven of your cousin by the dozen" method was enthusiastically promoted by the original Blue Zone discoverer Professor Michel Poulain, and our old friend Professor Jim Vaupel and a few coauthors, but everyone else's Blue Zone how-to guides mysteriously failed to mention this secret to extreme longevity.[4] Indeed, almost nobody noticed the idea at all, and the idea of Blue Zones looked set to fall into obscurity before an unlikely hero emerged to save, and sell, the concept.

Self-proclaimed award-winning journalist Dan Buettner's journalistic career began as an intern at NPR, the beloved US nonprofit news organization funded by charitable donations and grants. It was Buettner's duty to organize a celebrity croquet tournament for George Plimpton. While at

NPR, Buettner states he somehow obtained an "unlimited expense account" that he used to fund intern activities like riding around in limousines, eating at the finest restaurants, and staying in fancy hotels.*[5]

After this stint at NPR, Buettner began using his considerable marketing skills. He started to earn money by cycling around and telling people about it while taking sponsorships and selling books about his feats. These appear unironically impressive, with Buettner gaining three Guinness World Records for distance cycling while partnering with *National Geographic*. Or I should say he repeatedly claims to hold[6] three Guinness World Records, because the officials at Guinness disagree. After multiple inquiries, Guinness repeatedly stated only one team record ever existed for Mr. Buettner.†

As he cycled about, Buettner pushed the marketing process further by starting an online education company in which wee little kiddies click to decide his cycling adventures. He ended up selling the company for a "handsome sum" around 1997 or 1998‡ and, after living out the dream of a fifteen-year-old boy by getting paid to ride his sweet bike to cool places, seems to have done a few odd projects while enjoying the proceeds.[7]

Eventually, Buettner read an article about the growing baby boomer demographic, and "it occurred to him that these baby boomers, whose interests are shifting from the recreational drugs of their youth to prescription drugs like Prilosec [used to treat stomach ulcers] and Cialis [hypertension

* Buettner also states in the aptly titled *The Rake Magazine* that he squeezed out a deal where the tournament's sponsors would pay to fly him and several of his fellow organizers home every weekend. "But instead of saying that home was St. Paul [in Minnesota], where it was freezing, we rented an apartment in Freeport, Grand Bahama Island. So every weekend we'd get to fly to the Bahamas" on the nonprofit-funder's coin.

† Cycling lengthwise down Africa, one of his supposed records, was handed out "for the first time" several years after Buettner claimed to have done it: to the Tour d'Afrique team in 2003 and then to Ralph Tuijn (unofficially) and Robert Knol (officially) in 2011. When I asked a reporter acquaintance to nudge the Guinness people about the anomaly, I was informed that "we are only able to locate one record previously held by Dan Buettner in our database" (Guinness World Records, pers. comm.).

‡ Part of the sales package seems to have been a '90s video game called *MayaQuest: The Mystery Trail*—rating 6.5/10 out of four reviews on GameStop—that involved cycling about as Buettner to save the world from an asteroid impact. See "Mayaquest: The Mystery Trail," GameStop, https://web.archive.org/web/20230419143450/https://www.gamespot.com/games/mayaquest-the-mystery-trail/reviews/; The Rarest Gamer, "MayaQuest Trail (Windows game 1995)," YouTube video, 27:53, https://youtu.be/m_Ds1FMAJ-A.

and erectile dysfunction], would be interested in learning how to add a few years to their lives."[8] Buettner trademarked Blue Zones in the early to mid 2000s and set about discovering things. Or rather, paying people to discover things: After enlisting the help of his previous collaborator Professor Michel Poulain, the demographer who had formulated the inbreeding theory, Buettner landed some National Geographic Society grants and began to hire demographers to go find some more "zones." Buettner described the logic: "I made a deal with Michel [Poulain]. If he'd examine Rosero-Bixby's data and validate it, I'd find the money."[9] Buettner and his demographers christened two new Blue Zones and wrote up their results in a piece called "The Secrets of Living Longer" for *National Geographic*.[10] It became the magazine's third-highest selling issue of all time.[11]

In the process, some odd properties of the Blue Zones somehow became desirable traits. Inbreeding was lyrically transformed so that Sardinian populations were no longer genetically isolated with high inbreeding rates but were instead "genetic incubators, amplifying certain traits over generations."[12] Poverty became idolized as a collection of gentle physical activities that we should desire to emulate. An eighty-five-year-old's routine of milking cows, splitting wood, killing cattle, lifting rocks, and "walking and working my whole life, even when I don't feel great" was used to suggest that farming with no help or machines was an idyllic physical pursuit.[13] Even day-drinking homemade tub wine somehow became good for you, sold through beautiful photographs of elderly people.[14]

Encouraged by the warm reception of such lifestyle tips, Blue Zones was launched as a marketable lifestyle product, eventually backed by a smooth lineup of cookbooks and self-help guides.[15] Like the *National Geographic* story, the books were a wildly popular success, and the idea that extreme longevity is concentrated into certain geographic regions had been recrystallized in the public imagination.

Unfortunately, I have to say "recrystallized" because the idea was not exactly new.[16] Indeed, all of this was hauntingly familiar, especially to the editorial office of the National Geographic.

In the 1970s, Harvard professor of medical science Alexander Leaf explored the secrets of exotic populations that, it seemed, achieved remarkable longevity at a rate that outclassed the rest of the world. After great effort, expeditions, and extensive documentation, he had discovered the geographic regions of the earth in which the very longest-lived people resided. Professor Leaf's expeditions uncovered three regions of the world where people lived the longest and, like the scientists after him, somehow immediately understood what made these people survive to extraordinary ages and published the secrets in *National Geographic*.

These progenitor Blue Zones, called "longevity zones" back then, were three isolated regions in Soviet-era Georgian Abkhazia, Pakistan, and Ecuador.[17] Their familiar-sounding secrets to longevity? Mountains, low-fat vegetarian-ish diets, hard graft, poverty, and yogurt.[18] Notably, the yogurt claim was cultured by Danone to sell milky pots of apparent longevity—an invention that shot the company to profitability for the first time in years.[19] When asked about yogurt, a supposedly 121-year-old Georgian said people there never eat the stuff and instead began to reminisce fondly about when they were 28 and they stole their 12-year-old wife.[20]

Without laboring the point, let's just admit that none of these zones or secrets panned out, which is probably why these regions are not Blue Zones today. None of the Danone customers are 145 years old, and the lifestyle advice of "be positive and choose not to degenerate at age 65" was not exactly very useful.[21] They sold a lot of yogurt tubs, sure, but it all dissolved under scrutiny.

Take Ecuador's longevity zone in Vilcabamba. Documenting ages in Vilcabamba was initially complicated by "the destruction of early baptismal records in a church fire," but when a few independent scientists managed the feat, they did uncover a secret to longevity.[22] The people in these regions had an incredible talent for reaching old age, but the secret was not to live for longer. It was to do it faster.

Miguel Carpio Mendieta got older* this way: "when he was 61 in 1944 he reported an age of '70.' Five years later he was reputedly '80.' In 1970, at age 87, he was reputedly '121' and in 1974, at 91, he was '127' [*sic*]."[23] This aging approach was common, and by the time independent scientists were done with Professor Leaf's data, a 132-year-old had become 95, a 140-year-old was 96, and twenty-three centenarians had turned out to

---

*Admittedly, this is less alarming than his self-professed longevity secret: he was a professional hunter who washed in the blood of every animal he killed. (See Leaf & Launois, *Youth in Old Age*, 51.) This did not make it into the pro-vegetarian lifestyle advice.

be nowhere near 100 years old.[24] The average lifespan in Vilcabamba had, on reflection, fallen to be "similar to those for the US about 60 years ago [in 1922]."[25] I can't help but point out that the error distribution caused by these shenanigans also created a beautiful late-life plateau beginning around age 85, just like the original.[26]

As for the other regions, well, all had misreported ages and missing paperwork, the yogurt valley hardly merits debunking, and Pakistan's Hunza Valley suffered an outrageously large coincidence. The Hunza Valley had become a real longevity hotspot only *after* it was equated in the public imagination with the then wildly popular idea of Shangri-La, the idyllic valley where everyone lives past one hundred. Shangri-La is a work of fiction created by a Lancashire man who hadn't left the United Kingdom.[27] Yet, of all the millions of valleys in the world, Hunza became both a fictional and real valley of extraordinary longevity, with the fiction part coming first.

Incredibly good fortune, eh?

Ecuador has now largely given up on the claim, Pakistan is mysteriously silent about the Hunza, and only the Abkhazian chamber of commerce still clings to the idea that their longevity secrets are real.[28] For everyone else, it all looks silly in hindsight and damaging to the credibility of geographers. That's why it was surprising when a new secret set of longevity zones were enthusiastically taken up by the National Geographic Society.

With these past failures in mind, it seems appropriate to look at the newest longevity zones and see if they endure scrutiny as well as Soviet YogurtVille and Shangri-La.

There are now six Blue Zones. The Seventh-day Adventist Blue Zone of suburban Loma Linda, California, sits on the drive between Buettner's former home and Las Vegas, nested below the hills just south of the Costco in San Bernardino. The next, discovered after Buettner used a National Geographic grant to pay a Costa Rican demographer called Rosero-Bixby to look for longevity zones,[29] was duly discovered in one or (if you ask the demographer) two of the poorest and most remote regions of Costa Rica.[30] The other candidate Blue Zones are four islands: Sardinia in Italy, Okinawa in Japan, Ikaria in Greece, and the latest addition of Singapore.

The primary Blue Zone claim is that people in these regions survive past age one hundred at a world-beating rate through outstandingly high late-life survival.[31] Added to this primary claim are nine secondary claims about what causes this extraordinary longevity: a secret "power nine" list of behaviors to supposedly make you live to one hundred.[32] Fortunately, both the primary claims about longevity and the secondary claims about the causes are all directly and empirically testable. Let's look at the primary claims first, because if they are not true, the whole thing falls over.

In a move that seems to have utterly surprised the Blue Zones scientists, the collective governments of these countries seem to both care about and measure the longevity of their citizens. For example, from 1990 to today, the twelve to twenty-eight collective governments of the European Union (EU) have measured the average and late-life survival of their people. These governments have even compared the late-life survival—the average number of extra years you can expect to live after you reach eighty-five years old—across hundreds of regions in the EU. Where in this league table of late-life survival do the two European Blue Zones sit?

In 1990 Sardinia was tied 51st in Europe for old-age survival, out of 128 regions. The region containing Ikaria was 109th out of 128. These numbers have not really improved. In the latest rankings for 2021, Sardinia was tied 36th–44th, and the Ikarian region was tied 56th–65th. That is, these regions are not remarkable outliers for late-life survival globally: they are not even remarkable outliers locally. No Blue Zone regions ever ranked among the longest-lived regions of Europe. Instead, in 1990 the best region for late-life survival was downtown Brussels. In 2021 it was the picturesque Ticino, a quiet wedge of Switzerland between Lake Como and Lake Maggiore. These are not isolated hill villages filled with salt-of-the-earth shepherds but overtly rich and boringly safe places with good government infrastructure and free healthcare.

Reinforcing this picture, the Greek data also contain direct counts of those aged one hundred and older in the census, just after most of them were discovered to be dead and receiving pensions. According to EU data, once these errors were corrected, the administrative region containing Ikaria—which includes the threefold larger population of Samos—only had thirteen centenarians left in total and was nowhere near the top-ranked Greek region for centenarians per capita.[33]

Not exactly inspiring stuff. But what of the other regions?

The Centers for Disease Control and Prevention (CDC) evaluates average life expectancy in every neighborhood of the United States. Loma Linda, which occupies four census tracts, is in the 27th–75th percentiles of longevity[34] in the country.* That is, at least a quarter of US neighborhoods are estimated to live longer than this community. There are also more than 24 million Seventh-day Adventists living outside Loma Linda who are guided by the same cultural practices and ideals but whose communities somehow don't get their own Blue Zone.

We covered Japan already, but it bears repeating. The Blue Zone study was conducted around 2005. At the time, Okinawa was reported by the Japanese government as having more centenarians per capita of the forty-seven Japanese prefectures. Then, in 2010 at least 82 percent of Japanese centenarians turned out to be imaginary or dead.[35] After the ensuing and partially complete error correction exercises, Okinawa is now somewhere around 26th of forty-seven for late-life survival in Japan.[36] Locally unremarkable, once again.

Initially, the Costa Rican Blue Zone also looked okay until a tragic reassessment of the data revealed that at least 42 percent of the 99-plus-year-olds in Costa Rica were lying about or overstating their age.[37] Once the errors were corrected, old-age survival rates in the country fell to "near the bottom of the range," or "near the middle of the pack," with data that "produce a suspicious age pattern of mortality . . . casting doubt on the quality of national demographic data."[38] One study concludes that "we cannot draw a definitive conclusion regarding the reliability of the [old age] estimates because the evidence is circumstantial" and that "only in a context of improved reliability of the basic national statistics will we be able to determine with certainty the level of life expectancy in Costa Rica."[39] However, the Costa Rican demographer was unperturbed: the larger and more famous of his initial two proposed Costa Rican Blue Zones shrank by 90 percent, and the smaller one disappeared entirely before a new "secondary hotspot" was found at the other end of the country next to sunny Nicaragua.[40] Problem solved.

If you ask the people paid to find longevity zones, these regions have remarkable late-life survival. However, if you ask the independent governments or nongovernmental organizations that are not paid consultancies by people selling cookbooks, these regions have had unremarkable late-life longevity before, during, and after the Blue Zones studies.[41]

---

*Loma Linda is not a standard census tract but rather four neighborhood regions nested in the larger San Bernardino County, which of course has an unremarkable life expectancy, so this 27th–75th percentile range represents the four regions that together make up Loma Linda.

This brings us to Singapore. The Singapore Blue Zone was discovered last, when Buettner decided to find another Blue Zone region and reached out to Professor Michel Poulain. Professor Poulain is the original proponent of the Sardinia inbreeding theory who coined the term Blue Zone in the first place, and "Poulain said Buettner recently asked him to identify additional longevity zones in countries with highly marketable cuisines—India, Mexico and Thailand . . . 'cuisines that Americans love best.'"[42] When Poulain responded that there were no data to support such a designation, Buettner identified a sixth Blue Zone in Singapore: a global leader for marketable and delicious food, and a melting pot of Indian, Chinese, and Southeast Asian cuisines.

Buettner unveiled Singapore as the new Blue Zone in *Living to 100: Secrets of the Blue Zones*, with the country apparently "'engineered' as a society that lives to 100."[43] To me, this was somewhat hard to fathom given that Singapore did not have birth certificates one hundred years ago.* Universal birth certificates were first mandated in 1938—with a rollout somewhat complicated by the destruction and occupation of the nation in 1942 by the Imperial Japanese Army—leaving us with a rather funny unsolved mystery à la *Scooby-Doo*.[44]

How exactly does anyone know the "Living to 100" secrets of people in Singapore when the resident population did not have birth certificates until eighty-six years ago? How would you know who lived to one hundred, let alone compare the population-wide rate of reaching one hundred to the rest of the world?

The answer was, apparently, to utterly change the way that Blue Zones were defined *just* for Singapore. Instead of "Living to 100," the Singapore-only longevity measure used to label it a Blue Zone is an obscure public health metric called the Health-Adjusted Life Expectancy index, an index where Singapore came first in one lab's estimate for the latest year.

Many Singaporeans are skeptical. One health blogger reacted by asking "Where did you read this fake news?" before describing Singapore's stress-filled lifestyle and "salty and oily" diets with the "same nutritional profile as fast food."[45] These responses prompted Buettner to offer the grumpy nonanswer of "Singapore . . . that's where the excitement is. . . . The Saul

*Unless you were British or born to a British officer or merchant. The empire seems to have required paperwork if you were shagging the locals: seemingly, everyone else literally did not count on the British Empire register.

Newmans of the world can find ways to poke holes, but it's not where I want to spend my time."[46] I was glad to discover there were more Saul Newmans, but it hardly answered the questions.*

The Singaporean Minister of Health then gave an endorsement that raised the bar for irony, emphasizing the concept's complete disconnection from reality: "Singapore is distinctively different from the other Blue Zones. Good healthy habits are not inherent in our traditional culture and lifestyle. Singapore is nothing like Okinawa or Sardinia. On the contrary, we eat food that is rich in sugar, salt and santan [coconut milk], and a lot of our food is deep fried; we are not endowed with a vast expanse of nature that encourages outdoor activities; the pace of life is fast and stressful and many families keep to ourselves and rarely even talk to our neighbors. So we have the opposite attributes of the other five Blue Zones . . . we do not have a culture of healthy lifestyle."[47] In other words, the Singaporeans live a long time by being completely *unlike* the Blue Zones in every way.

This Singapore switch brings us neatly to the secondary lifestyle-based claims: those nine "secret" causes of remarkable late-life survival—the keys to living past one hundred. Rather than trying to exhaustively—or exhaustingly—detail the full list for all six zones and all nine claims, let's look at the single best-documented zone of Okinawa and go from there.

Okinawa is subject to the meticulous recordkeeping of the Japanese government. Japan has direct comprehensive surveys of nutrition, behavior, longevity, and seemingly everything else it is possible to measure, often going back to 1970, for every single prefecture. These surveys are routinely representative of over 96 percent of Japanese households, happen virtually every year, and the data are published in English.[48] While there is no indication that most scientists involved in the Blue Zones or aging research have read or cited these studies, they provide data on virtually every claimed "secret" for longevity.

To save a lengthy discussion, I will just quote the claim and then cite the independent Japanese government data measuring that claim. I will also make the text **bold** to emphasize the measurable stuff.

---

*I have only found one: a professor of insurrectionist politics, whose students (alarmingly) began to email me unsolicited questions about strategies for overthrowing "the West" and what I thought of Taliban tactics, before I figured out I had a doppel-namer.

Claim

"**Move Naturally** . . . The world's longest-lived people don't pump iron, run marathons or join gyms. Instead, they live in environments that constantly nudge them into **moving without thinking** about it. They **grow gardens** and don't have mechanical conveniences for house and yard work."[49]

Evidence

Okinawans have the worst Body Mass Index (BMI) in Japan.[50] When compared to every other population in the country aged seventy-five and older, the oldest Okinawans also have the worst BMI in Japan.[51] Okinawans have held both records from 1975 onward, when the cohorts aged seventy-five and older were all born in 1900 or earlier, so this is a century-spanning trend.[52] Okinawans rank second to last of forty-seven prefectures for engaging in rest and relaxation time.[53] Even gardening is measured, and Okinawans rank third to last, only just beating the largely yard- and garden-free megacities of Osaka and Tokyo.[54]

Claim

"**Plant slant . . . Beans**, including fava, black, soy and lentils, are the cornerstone of most centenarian diets. **Meat—mostly pork**—is eaten on average **only five times per month**. Serving sizes are **3–4 oz**, about the size of a deck of cards."[55] Note that eating 3–4 oz servings "only five times per month" is equal to **5.1–6.8 kg of meat a year**.[56] In addition, this idea has been reinforced by the repeated claims that Blue Zones diets are plant based and Okinawans supposedly eat lots of sweet potato as a longevity booster.[57]

Evidence

Okinawans eat 19.1 kg per year of pork, 6.2 kg per year of beef, 12 kg per year of chicken, and 2.6 kg per year of "other raw meats," for a total of at least 40 kg of meat a year, with an additional 20.5 kg of seafood consumption.[58] This places them firmly in the middle of the pack for Japanese meat consumption and way above the global average per capita meat consumption of 33.9 kg a year.[59] The nutritional composition of this meat is also somewhat questionable: Okinawa consumes over 95 percent of all the (heavily processed) lunchmeats in Japan and has a particular soft spot for Spam—and ranks first of forty-seven prefectures for Kentucky Fried Chicken consumption.[60] Incidentally, given the claimed health benefits of oily fish, Okinawa ranks dead last for seafood and oily fish consumption in Japan.[61] None of this is new: it is a consistent trend from the start of records in the 1970s until now.

Okinawans are last out of forty-seven prefectures for sweet potato consumption and taro consumption.[62] Although beans are not assessed as a single dietary category, Okinawa ranks 44th (fourth lowest) for string bean consumption, 44th for soybean paste consumption, and second to last (46th of 47) for bean curd consumption.[63] As for plants in general, Okinawans are last out of forty-seven prefectures for consumption of root vegetables, last for consumption of pickled vegetables, 39th out of forty-seven prefectures for salad consumption, and last for consumption of leafy vegetables.[64] Again, these are not new or recent trends, and they really should eat some more greens.

Claim

"**Belong** . . . All but five of the 263 centenarians we interviewed **belonged to some faith-based community**. Denomination doesn't seem to matter. Research shows that attending faith-based services four times per month will add 4–14 years of life expectancy."[65]

Evidence

Japan is a highly atheist country. Okinawans are 93.4 percent atheist, the most atheist prefecture in Japan,[66] and spend the third least amount of time in religious ceremonies.[67]

Claim

"**Purpose** . . . The Okinawans call it 'Ikigai' and the Nicoyans call it 'plan de vida'; for both it translates to 'why I wake up in the morning.' Knowing your sense of purpose is worth up to seven years of extra life expectancy."[68]

Evidence

Japan struggles with high suicide rates nationally, and Okinawa has the ninth-highest suicide rate in Japan.[69] Older Okinawans do not fare better. In 2019 Okinawa had the fourth-highest suicide rate in Japan for people aged sixty-five and over.[70]

Claim

"**Loved Ones First** . . . Successful centenarians in the blue zones put their **families first**. This means **keeping aging parents and grandparents nearby**

or in the home (It lowers disease and mortality rates of children in the home too.). They **commit to a life partner** (which can add up to 3 years of life expectancy) and **invest in their children** with time and love (**They'll be more likely to care for you** when the time comes)."[71]

Evidence

Okinawans are less likely to cohabit with older people. They have the second-lowest percentage of households with an inhabitant aged sixty-five and older in Japan.[72] As for committing to a life partner, Okinawans have the second-lowest percentage of people who are married.[73] The Okinawan divorce rate has exceeded the national average every single measurable[74] year since 1911. Okinawans have the fifth-largest fraction of single parent households in the country and, among older Okinawans, the highest proportion of divorced men aged sixty-five and older and the fourteenth-highest proportion of divorced women in that age group nationwide.[75]

Claim

"**Wine at 5** . . . People in all blue zones (except Adventists) drink alcohol **moderately** and regularly. **Moderate** drinkers outlive nondrinkers. The trick is to drink **1–2 glasses per day** (preferably Sardinian Cannonau **wine**), with friends and/or with food. And no, you can't save up all week and have 14 drinks on Saturday."[76]

Evidence

Okinawans drink regularly but not "moderately" and not "mostly wine": they drink everything and mostly beer. Okinawans rank second for beer consumption nationally (61.0 liters per year), eighth for wine consumption (3.6 liters per year), and fourth for total alcohol consumption (8.8 liters per year of pure alcohol).[77]

However, the core evidence-based problem for this claim is not the numbers. It is the volume of alcohol Buettner and the Blue Zones are recommending. Two glasses of wine a day is twice the National Health Service guidelines, and far above CDC guidelines, for "heavy drinking." No physician in their right mind would recommend their patients drink every day. It is a recipe for alcoholism, promoted as a cure by a man with no medical qualifications. I am an Australian from the edge of the outback, and even I understand that drinking is not good for you.[78] It is unfathomable to me that this claim was taken seriously for twenty years, and treated with complicit silence, by physicians and scientists. If I were an American and I suffered harm by following this guideline, I'd be reaching for my lawyer.

Claim

"**Right Tribe** . . . The world's longest-lived people chose—or were born into—social circles that supported healthy behaviors, Okinawans created 'moais'—groups of five friends that committed to each other for life. Research from the Framingham Studies shows that **smoking**, **obesity**, **happiness**, and even **loneliness** are contagious. So, the **social networks** of long-lived people have favorably shaped their health behaviors."[79]

Evidence

Sadly, this claim is so vague as to be virtually impossible to test. How exactly can a *tribe* be defined, in a single testable framework, across modern Japanese, Greek, US, Italian, Singaporean, and Costa Rican societies? That said, we can test the patterns of the supposedly "contagious" health behaviors of "**smoking**, **obesity**, **happiness**, and even **loneliness**."[80] Some 37.4 percent of male and 9.7 percent of female Okinawans smoke, both above the current global average and far above the smoking rates in other rich countries.[81] We covered obesity before—Okinawa has the highest BMI in the country every year—and happiness was rather grimly shown in the suicide rates.[82] As for loneliness, we only have measures of old-age diagnosed clinical depression, which often depends more on healthcare access (which is poor) than anything else. The rate of Okinawan patients over sixty-five diagnosed with major depressive disorders was the only indicator that was even slightly "good," ranking Okinawans as seventh.[83]

Similar laundry lists of problems arise in all the other Blue Zones, but to avoid the risk of tedium, let us skim through to the fun ones. First, let's see what got left out of the lifestyle tips.

The single most famous Sardinian food before Buettner turned up was *casu marzu*, a cheese infested with live maggots, which the EU banned because the maggots were routinely surviving the digestive process and eating the intestines of the unlucky Sardinians. The "cheese that bites you back" did not somehow make it into the cookbooks. Neither did other problematic food and lifestyle exposures. For example, where were all the Okinawan Spam recipes?

They also skipped a much darker error-generating process. In WWII, the US Army had cataclysmically firebombed and invaded Okinawa, killing tens of thousands of civilians and burning entire cities, including the birth and death records. After the war, this paperwork was replaced by the US Army government based on self-reported ages, by GIs and officers who generally did not speak Japanese, in a calendar that the Japanese citizens did not use. Unsurprisingly, this all led to errors, with the result that you can predict 79 percent of the variation in where Okinawan centenarians live just by mapping American bombing campaigns.[84] Another secret to longevity that somehow missed the cut-off.

The historical explanations in Greece were older: the island of Ikaria was famous for its "immortal waters" of longevity all the way back to 500 BC. Why did this original Ikarian secret to longevity not make the cut? Perhaps because the springs are "highly radioactive" and contain extraordinary doses of uranium, thorium salts, and radon.[85] This radiation builds up in the soil and water of Ikaria to such a degree that despite Ikaria being just 0.17 percent of the Greek landmass, the island has triple the maximum dose of gamma radiation observed anywhere else in Greece.[86] Ikarians are exposed to these elements at baseline levels that impart almost double the maximum safe dose of gamma radiation during everyday activities.[87] If you also drink and bathe in the spring waters, which many Ikarians do all the time, you'll risk roughly triple the maximum safe dose of radiation.[88]

Floating in these warm glow–inducing "waters of immortality" are another more modern additive. When the Greek Civil War ended in 1949, over thirteen thousand die-hard communist soldiers and partisans were exiled as punishment to Ikaria—which only had eight thousand occupants left after the Nazis had starved the place—and many exiles stayed there permanently.[89] In a country already partial to systemic pension fraud, the tiny rocky island became populated by people who hated the government and believed in the equal redistribution of wealth.[90] And 100-plus-year-olds, apparently.

The island of angry, radioactive communists.

Sadly—like the firebombings, incest, and maggot cheese eating—this lifestyle never made it into the cookbooks either.

The general disaster of the supposed Blue Zone lifestyle can, however, be best summarized by just one finding. According to the government of Japan, where do Okinawans rank for eating sweet potato, including the gloriously iconic purple variety pictured on the front of the Blue Zones Netflix documentary *Living to 100*? Dead last. By a mile. Every year. Forty-seventh out of forty-seven prefectures.

If you ask the Okinawans what they eat, and the Japanese government do it intensively, the Okinawan diet is meat-heavy, vegetable-absent, relatively beanless, and utterly unlike the characterization proposed in the cookbooks. These completely contradictory dietary patterns, observed by the Japanese government through massive nutritional and household surveys, are not a recent trend but a long-lasting cultural pattern that extends all the way back to the beginning of government records. We know this because the Japanese government asked citizens through comprehensive surveys representing 96 percent of all households in the country.

This leaves the question: Where on earth did the Blue Zone diet come from?

The Blue Zones claim to have interviewed 263 centenarians, but they do not seem to have published the interviews.* Given what we have seen from the heavy-smoking, heavy-drinking lives of the world's supercentenarians, their lifestyle tips probably resemble Overton's catfish and cigars or Calment's chocolate and cigarettes more than any kind of salad promotion.

Naturally, when I asked all these questions out loud, I received no response. Instead, the excuses began.

Apparently, the ever-so-recent Blue Zones began suddenly vanishing—just after the criticism and data had arrived.† The largest Costa Rican zone shrank to a tenth of its original size, the smaller one proposed by the demographer disappeared, and another "secondary hotspot" was discovered near the Nicaraguan border—all in the face of a vaguely defined "vanishing advantage" and "cohort shift" of unknown origin.[91] The Sardinian Blue Zone also wandered off a bit, and is now on two sides of the island. Okinawa was even worse, evaporating completely in the face of the great evil of "Western diets," whatever they are, to the point where "even though yes, there are still some leftover 100-year-olds from the traditional way of living, when

*It is not even stated if these interviews are population-representative or random samples, or why we should ignore the gargantuan studies conducted by a government in favor of a few hundred "trust me bro, these beans are the secret" chats that have never been reported in full.

†Because it was added fifteen years after the rest, it was not possible to claim that the Singapore Blue Zone had "just left." So, it seems to have received a special goalpost-shifting treatment. While the rest of the Blue Zones were defined by a high rate of survival to extreme old age, Buettner redefined Singapore's Blue Zone status so that it rested on a claimed high average healthy-life expectancy.

they're dead, the next cohorts will not be long-lived."[92] All of the Blue Zones were apparently degrading in the face of the greatest of dietary evils: behaving and eating like Americans, which the eighty-plus-year-old Italian, Greek, and Japanese people were apparently busy doing.*

This looks extremely disingenuous to me. Buettner claims that Okinawa was healthy in the early 2000s when he began his study and has now degraded. But the health metrics from long before, during, and long after the study place Okinawa at the bottom of the pack. The data start in the 1970s when the first comparative health studies of Okinawa begin and, as we saw above, these exhaustive government efforts uniformly contradict everything the Blue Zones scientists propose. The same happens in Ikaria and Sardinia, which have mediocre late-life survival from at least 1990 until 2021 and every single year in between.

Even more damning, the continuous updating of independent data show that health in Sardinia, Ikaria, and Okinawa has been rapidly *getting better*. As their diets supposedly "Westernize" and their Blue Zone behaviors reportedly disappear, the European Blue Zones have been increasing their over-eighty-five life expectancy and marginally catching up to the rest of Europe. This is not some niche survey but a finding generated by collective EU governments across 36 years. Okinawa has also been improving and is also rapidly catching up with the rest of Japan as their diets and behaviors "Westernize"—whatever this is supposed to mean.[93] The Okinawan BMI data are a great indication of this trend. The BMI has been continuously measured in Okinawans older than seventy since 1975: that is, the first measurement was for everyone born in 1900 or before. These estimates have been updated continuously. By this metric, Okinawans were *always* the least healthy population in Japan from 1975 until 2018. During this period, BMI increased across Japan, but Okinawan BMI increased at a slower rate than the rest of the country until they finally closed the gap with the next-least-healthy prefecture in 2018.[94] All as the Okinawan diet was apparently Westernizing.

The same picture is playing out in life expectancy, both in Okinawa and in the other Blue Zones. All of them are improving from relatively poor starting positions. Nothing degraded; it was just never there.

---

*Not, say, adapting their own rich and independent food cultures to an ever-changing and increasingly diverse cross-cultural palate, adapted to suit their own tastes. Nah, it must be Western imperialist hamburgers or whatever idiotic slop is supposed to outclass the local *tajarin al tartufo*.

Finally, we reach the only Blue Zone that cannot Westernize its diet—Loma Linda, California, in the United States. Faced with evidence that the Loma Linda Blue Zone longevity was contradicted by CDC data, Buettner revealed to *The New York Times* that Loma Linda was "kind of an outlier" that was only "included because his editor at National Geographic told him 'you need to find America's blue zone.'"[95] Buettner then stated that he "never bothered to delist [Loma Linda]" as a Blue Zone during almost two decades of sales and marketing. Loma Linda is even still being promoted as a Blue Zone at the time of publication.

Loma Linda was discovered on the dusty main road from Buettner's former home city of Los Angeles to Las Vegas, after his editor told him to "find" a place. This raises some rather awkward questions about the scientific integrity of everyone involved. The affair also raises some questions for Loma Linda–based Adventist Health, a religious nonprofit that acquired Blue Zones LLC in 2020 for a reported $78 million USD, given Adventists are supposed to stay devoutly sober and Blue Zones promoted drinking every day.[96] Yet the editorial nature of the "discovery" process has not prevented leading demographic experts, such as Craig and Bradley Willcox, from promoting Loma Linda as equal to other longevity zones and selling their own would-be wonder diets.[97]*

None of these revelations matter much, sadly.

Sharecare Inc. purchased the Population Health business of Healthways, including Buettner's "Blue Zones Project," in 2016. Blue Zones Project uses the Blue Zones trademark[98] under license from Blue Zones LLC (which was, in turn, owned by Adventist Health).[99] The relevance of this slightly complicated and boring corporate structure, with Sharecare owning Blue Zones Project and Blue Zones LLC owning the trademark, would become apparent after I started asking thorny questions in 2019.

For the uninitiated, Sharecare is a company cofounded by one Dr. Mehmet Oz, a Harvard-educated physician who used to be famous for talking on the *Oprah Winfrey Show*. Dr. Mehmet Oz is the proud three-time winner of the

---

*As recently as 2023, Buettner was promoting Loma Linda in *Fortune* as a literal paradise, where "like in the Garden of Eden, they eat foods the Bible praises, like the 'seed-bearing plant' from Genesis 1:29." For what it's worth, virtually all plants bear seeds, and having been in Loma Linda, it looks to me more like Barstow than paradise.

Pigasus Award* for pseudoscience—in 2009 for promoting "such quackery as Reiki," in 2010 for promoting pseudoscience and "quack theories that have no scientific basis," and in 2012 in the "refusing to Face Reality" category for his "continued promotion of quack medical practices . . . [including] pseudoscientific Reparative Therapy to 'cure' gay people."[100] This track record led to him being (eventually) removed from Columbia University after a group of doctors decried his "disdain for science and evidence-based medicine [and] egregious lack of integrity by promoting quack treatments and cures in the interest of personal financial gain" before calling him "a menace to public health."[101] He was, naturally, incoming president Trump's first choice to lead Medicare and Medicaid.

Sharecare has proven to be a perfect buyer of the Blue Zones Project. By 2024, the new owners, who pay licensing and royalty fees to use the Blue Zone name and tenets, had begun selling certificates town to town, allowing cities and small towns everywhere to become "Blue Zone Certified." Although many of these deals are sealed by non-disclosure agreements, *The New York Times* found that a surprisingly large number of towns and cities paid between $3 and $40 million to participate. New York State even announced a $1.2 billion development of 2,600 apartments in Brooklyn that will include (possibly unofficial) Blue Zone–inspired features.[102] Personally, I look forward to the part where they release the geriatric radioactive communists and give out the wriggly maggot cheese to the Brooklynites.

The subscription costs for becoming a Blue Zone are so high that after nonprofit Wellmark Blue Cross and Blue Shield—which had invested over $25 million to support towns in the state of Iowa to participate in a Blue Zones initiative—decided not to renew the contract, numerous towns opted out, citing the continued costs they would need to pay to continue to use the brand name.[103] After this arrangement ended, the ability to provide Blue Zone–style insights was undertaken by a Wellmark nonprofit initiative, Healthy Hometown, which began doing it all for free. During its time as a Blue Zone adherent, Iowa went from being ranked as the 16th-healthiest state to the 21st-healthiest state in the United States. In Buettner's words, this short, sharp drop in the rankings apparently happened because the program had "not been funded

* A prestigious award, consisting of a psychically bent spoon resting on a "flimsy and transparent" plastic base, the Pigasus Awards are given to prominent claims that will be true when pigs fly (i.e., never). According to the wonderful organizer, "winners are notified telepathically and are allowed to predict their victory in advance." Oprah Winfrey shares the award, but Dr. Oz seems to be the only three-time winner.

there for long enough to make the initiatives stick"[104]—even though another Blue Zone town called Brevard in North Carolina was claimed as an outrageous success after three years. In Brevard, however, it was the "Blue Zones Project" operator Sharecare that was measuring the outcomes.[105]

In return for these huge sums, the public apparently receive valuable insights into things that are strikingly obvious to any public health official with a quarter of a brain ("we should nudge people to exercise"), are of extremely dubious health value (like encouraging "faith-based activities"), or are painfully bad for you (like daily drinking). As a result, the company is taking in huge amounts of public money, often without their results being independently evaluated to see if they even work.[106]

Normal public health programs must be evaluated for effectiveness against other programs that otherwise would have used the money. For example, these programs should outperform regular public health programs run by qualified public health professionals. These evaluations never happened.[107] Instead, the first trials of the Blue Zone certification process were primarily judged using data from Sharecare.[108] They promptly marked their own homework and gave themselves an A+.

Basic safety steps seem to be missing. To prevent some random company from just handing out thalidomide, energy drinks, or peanut butter on the basis that they think it helps public health, public health interventions in the United States are legally required to clear an independent ethics board.[109] This is done to make sure that the statistical design of the trial adheres to the Declaration of Helsinki, makes sense and is specified ahead of time, reports sample sizes and success metrics, is assessed against a business-as-usual scenario of "What if we spent this money on normal public health interventions?," and ensures that nobody in the public is placed at risk. An independent ethics review board would prevent a company placing the public at risk by, for example, taking a bunch of public money and telling everyone to drink daily without any statistical rigor or plan.[110]

There does not seem to be a legally mandated ethics review available for any of the Blue Zones interventions, and none of the work reports any data or preregistered (preplanned) statistical design.[111] Although a standard health intervention would report them, I have not seen any informed consent forms, nor any indication that people involved have the capacity to opt out. There seems to be no published plan or analysis at all for an intervention, and on the few occasions a paper has been released after the fact, the "paper" did not have a methods section.[112] The main reported study only has four citations, two of which refer to Blue Zones cookbooks.[113]

So how do we know this program is safe or if it even works? Well, instead of clinical best practices designed to protect the public, Blue Zones are assessed by the polling organization Gallup. At least, that's what it seems like at a casual glance. When asked about efficacy, Blue Zones employees point to "Gallup surveys" that suggest their programs are improving health and lifestyles, with rosy conclusions like "Based on Gallup's surveys . . . well-being *does* improve after blue zones are established" or the gem that "it's very rare you get the opportunity to address community health issues directly and utilize a world-class measurement tool—the [Gallup-Healthways] Well-Being Index—to chart your progress, but Blue Zones Project made it possible."[114]

In 2024 a Gallup pollster, contracted to assess the Blue Zones program, revealed the reality that "improvement measurements from the Blue Zones Project come from the Blue Zones team."[115] This is because the Blue Zones Project's parent company Sharecare bought and owns the Gallup-Healthways Well-Being Index—the metric used to assess the Blue Zones interventions. They bought the Gallup index at the same time as they purchased the Blue Zones Project in 2016 and use the index to measure whether their own product is working.[116] They are scoring the benefits of their own program.

One of the latest grand plans for the Blue Zones movement is for the American College of Lifestyle Medicine to certify doctors as a "Blue Zone Certified Physician" under the "Blue Zones Health" initiative, a partnership of Blue Zones LLC and a company called Healthly.[117] Again, the reality of marking their own homework is not mentioned. Joe Daly, the managing partner at Gallup in charge of the well-being division, instead states that "the launch of Blue Zones Health is incredibly exciting. For years, Gallup research has highlighted the impact of personal and community well-being on one's health . . . Combining the research Gallup is conducting with Blue Zones . . . has the potential to make a massive impact on well-being."[118] At no point does he mention that Sharecare both runs the Blue Zones Project and owns the Gallup-Healthways Well-Being Index.

As a result, nobody appears to have independently measured the success or failure of the program or if the Blue Zones programs even work.

It hardly matters. The profit has been made. They got their money, and the public will never get it back.

7 Oxford Without Answers

As the Blue Zone drama and the Jeanne Calment fight began to play out, I went through a change in personal circumstances. By early 2020, the national Australian science budget had been slashed to more than forty-year lows by successive governments. They reached a shocking nadir when the pandemic began. Australia locked her borders, saving tens of thousands of lives but cutting off the flow of international students. This throttled the only revenue keeping the nation's universities alive. To make matters worse, a law that was passed to support businesses affected by the pandemic was deliberately rewritten, by a conservative government, to exclude university staff and scientists. Roughly 20 to 25 percent of research staff lost their jobs, including many health scientists and virologists, at the start of the pandemic.

I was fortunate enough to make an escape and got a job as a research fellow at the University of Oxford, at the Leverhulme Centre for Demographic Science, under a boss who assured everyone that they were ready to "disrupt and realign" demography. It sounded promising and I was excited. When I started work in early 2021, I thought I could make a great deal of difference in a field that was suffering a great deal of harm. Misinformation, misconduct, and outright fraud seemed to be dominating aging research. I had reasoned that if people could just see the behavior and patterns I had seen, backed up with evidence and code, things could begin to change for the better. I felt like if I could just hit the toxic fiefdoms hard enough, I might introduce some small measure of accountability.

I was making a very stupid assumption.

The Blue Zones, the most popular concept in demography for the past two decades, were pulled apart faster than spare ribs in a US Army mess hall.

All this did was underline the size of the problem and embarrass powerful people. For twenty years, a wide array of demographers, epidemiologists, geographers, and medical scientists had not noticed the problems or, worse, noticed and stayed silent.

Confronted with what I had discovered, those not personally involved in the case laughed their asses off, those who had backed the idea seethed in silence and lobbed emails in private, and everyone else continued to do nothing.

Inevitably, I engaged in a lot of academic bun-fights behind the scenes.

I had pointed out that even careful validated data had repeatedly turned out to be 100 percent errors. A common (private) response was to suggest that this could not happen because extreme-age data had been "carefully validated." When I repeated myself, again pointing out that "carefully validated" data had repeatedly turned out to be 100 percent errors, over and over, the response was to repeat the same reply: yes, but our data were *carefully* validated. The most careful validations ever performed* did not work → yes but *our* validation was too careful to allow errors → the most careful validations ever performed did not work → yes but *our* validation was too careful to allow errors. Around and around the merry-go-round of logic we went.

One of my favorite pieces of bad logic is the question people ask in movies or TV when they must discover which one, of a superficially identical pair of people, is the evil robot, clone, or twin. When they're pointing the gun and asking Steve or SteveRobot for proof of their identity they say, "Tell me something only the *real* Steve would know."† The hidden-in-the-open Yogi Berra logic is that if only the *real* Steve knows something, the person asking the question will *not* know it by definition because they are not the *real* Steve.

---

*Even when restricting myself to validated "world's oldest person ever" titleholders, I could list the validated and then debunked cases of Carrie White, Charlie Smith, Mary Duckworth, Walter Williams, Arthur Reed, Shigechiyo Izumi, Jesus Arenas, Sara Corzo, Martha Graham, Lucy Hannah, and Pierre Joubert—the man whose record took a century to debunk. These people passed the most intensive case validation possible using modern methods, and all were completely fake.

†What they mean is that they want them to tell them something only *we* would know, but the line has become a cliché and never seems to bow down to my petty desire for pedantry.

Delightfully, this circle of bad logic is also a common technique for demographers to validate the age or identity of older people. If a demographer wants to know if somebody is 110, they ask them something only a 110-year-old would know.[1] Seriously. That's the method. I often see it in the wild. Professor Robine had a beautiful example when he defended his validation of Jeanne Calment, by confidently stating, "We had information that only she (Jeanne) could have known, such as the name of her math teachers or the housekeepers in her building at the time. We asked her questions. Either she couldn't remember or she answered correctly, but her daughter couldn't have known that."[2]

Professor Robine is certain this knowledge could only be known to the *real* Jeanne Calment despite the fact that he knew it himself and he is Professor Robine.

Amusingly, nobody else ever seems to have spotted the logical circle despite some journalists spelling it out by accident.[3]

Then there is the second-order meta problem. Somehow, Professor Robine knew that Calment's daughter (or some other impostor) cannot know something that Jeanne knows despite the fact that he knows it himself. Despite the fact that this impostor/daughter must have lived in Calment's house, with all her as-yet-unburned personal effects and photos and diaries, and may have known Calment for years.

Most of this roundabout arguing was, regrettably, a private bit of fun: a kind of *Where's Waldo* of logical errors, hidden from public view.

One of the few public replies came from a French professor of demographer, Jacques Vallin. He had tried to explain why the very poorest parts of France, the overseas territories of Guadeloupe and Martinique, had the most 110-year-olds.[4] The explanation couldn't possibly be my theory of pension fraud, clerical errors, or bad governance. Instead, Vallin decided, "In my view, only a genetic hypothesis can explain such an extreme prevalence of supercentenarians: the tremendous health selection effect of slavery."[5]

It was a wholeheartedly racist approach. Not the "he said the N-word" kind of idiot-at-a-barbecue style of racism but the rarer kind, where someone publishes a scientific paper saying that one "race" was inherently more robust than another because of the benefits of dying as a slave. This "benefit" of "selection of the strongest individuals" was even amplified, in Vallin's view, by "low immigration and high fertility among black people compared to white."[6] It was sadly not a new idea but the extension of the long-unchallenged demographic theory, where "a consensus seems to have developed that the crossover [where underprivileged groups reach extreme ages more often] is indeed real and results from higher early mortality that removes less hardy blacks."[7]

The theory is about as enticing as a molten-lead martini. Like most racism, it was also trivially easy to debunk, even on its own terms. Mainland France has better recordkeeping, lower poverty rates, and longer average lifespans than Guadeloupe and Martinique. Mainland France also has more than 5.3 million descendants of the transatlantic slave trade,[8] about 8 percent of the total population.[9] None of them ever reached age 110.

The response of my colleagues to Vallin's arguments was not, however, one of shock or sadness. Instead, I was ignored by most, firmly told to shut up by many, and commiserated with by an otherwise silent few. Instead of receiving information, or even just acknowledgment, the aging science field was silent. This was often by design and, in a very British fashion, I soon found myself the subject of deliberate and crushing social isolation.

Not one to squander my quiet time, I began to spend my hours understanding: how did it happen? How deep did it go?

Age is a huge risk factor, often the largest risk factor, for virtually every nongenetic disease. Yet if a developmentally mature person arrives at a hospital without paperwork, there has been no machine in a hospital that measures age. You cannot place a human in a gigantic box, press a button, and wait for it to go "BING, they are 54 years old" in a delightful robot voice. As a result, one of the most fundamental and important metrics in medicine—age—has never been fully validated against a physical standard.

This excludes some rough exceptions. Newborn babies are an obvious one: you can validate their age with a stopwatch if you like, right up until the midwife whacks you with a drip stand and tells you to get out of the maternity ward. Developmentally immature people can also have their age estimated using their teeth, which emerge on a loose schedule, and their cartilage, which turns into bone on a fairly tight sequence, making it possible to approximately measure age using an x-ray machine. However, the last pieces of cartilage to turn into bone, and stop the clock, are the medial ends of your clavicles around age 26 and the crack just in front of the big hole at the base of your skull—or "basal suture immediately anterior to the foramen magnum," if you want to be all fancy.* The tissue in this crack turns to bone and welds the base of the skull in place around age twenty to twenty-one, the clavicles finish age 26. After that, we have nothing to tell us your age.

Instead, we have placed absolutely all our trust in a paperwork system that is, like every paperwork system everywhere, prone to manipulation and errors. It was clear that we were not detecting these errors. The systemic failure of demography and medicine to detect so many mistakes—hundreds of thousands of dead people had been turning up in 100-plus-year-old populations that had been "exhaustively" checked—did not point to an uplifting conclusion.

Some hints come from the United States, which makes a fascinating case study.

The latest 2020 US census shows a remarkable amount of "age heaping": the preference of some people to round off their age by a multiple of five or ten years because they don't know how old they are.[10] Tellingly, the amount of age heaping has been getting steadily worse since 2000 as more people forget or misreport their age or the ages of people in their household.[11] The Census Bureau paints a strangely rosy picture that only "5.0% to 9.9%" of adults—around ten to twenty million respondents—are rounding their ages off to a multiple of five. This is considered fine in demography but speaks volumes. Theoretically, old-age data become unusable above age ninety if you only suffer a 0.01 percent error rate. The United States had 100- to 500-fold higher error rates than this in the 2020 census.[12]

So much for the census data; now let's look at birth certificates. Most US states did not have any birth registration until 1900–1910, and the lower forty-eight states did not all have certificates before 1933.[13] The last two states to get basic registration were Hawaii and Alaska in 1959, meaning a Hawaiian or Alaskan over the age of sixty-five is very unlikely to have any direct record of their birth.[14]

This general absence of paperwork has led to some striking patterns. During WWII, Congress passed a law limiting participation in the wartime economy to people who could prove their citizenship. More than half of

---

*As an aside, if this happens too early, the brain is restricted: I wrote my first (not very good) thesis on the evolution of genes involved in timing the closure of skull sutures.

American citizens did not have any paperwork at all, let alone paperwork with their age written on it, and a larger majority of adults did not have a birth certificate.[15] So, with Congress not wanting to break its own laws, and needing wartime factories to run at warp speed while the Army grew at incredible pace, the US government began to simply dish out birth certificates based on whatever people told them. Tens of thousands of fully adult people were "born" each month, at whatever age they told the local registry or post office, without any real concern as to accuracy.[16]

The result was a lot of screw-ups, piled into an already-shambolic filing system.* As late as 1960, 40 percent of US citizens had different ages on their death certificates than the age they had self-reported on the census.[17] Over 27 percent of men and 57 percent of women wrote down inconsistent ages, like Richard Overton reporting his four or five different birth years, often because they had no idea what their age was anyway.[18] This error rate rose dramatically among the victims of racism and racist governance. Likely through a combination of fewer and worse educational opportunities, lower local government funding, and worse-kept government records, some 60 percent of African Americans had at least a one-year difference between the age on their death certificate and the age they had self-reported in the census.[19] Overall, between 12 percent (white male) and 42 percent (nonwhite female) of people had a five-plus year difference, and a staggering 8 to 11 percent of white and 24–30 percent of nonwhite people's official ages were off by a decade or more.[20] This was in 1960, for people about 50–70 years of age, those who, if they died now, would be 110-plus-year-old supercentenarians.

As a result, a huge number of the ages written on birth certificates—many dished out at speed in a local post office in wartime—are almost certainly incorrect. If you told the bloke at the post office you were fifty-six, fifty-six you were. As these birth certificates are official, with the ages copied forward to all later documents and used to replace, issue, or correct "earlier" documents, the incorrect dates routinely generate undetectable errors.

Added to this historical chaos is the modern problem of identity theft. It is trivially easy to buy a birth certificate. In the United States, over six thousand official entities issue over fourteen thousand different types of

* Two states didn't have any birth certificates at all, and many had only rolled out birth certificates twenty to thirty years before.

official birth certificate, many of which can be legally purchased online as a brand-new birth certificate without any identification from twenty-four states. According to the US government, "fraudulent birth certificates are easy to obtain" and identity theft is common, and as a result the government has long advised against the use of birth certificates for proof of identity or age—which is, of course, exactly what epidemiologists, scientists, demographers, and doctors have been doing.[21]

After one recent investigation into birth certificate fraud, the US Office of the Inspector General stated that "a birth certificate is proof only that a birth occurred and was recorded. . . . However, birth certificates are used as proof of age, citizenship, and identity."[22] It goes on to state that "because a birth certificate cannot be positively linked with an individual, their use for that purpose is questionable."[23] Birth certificates are, nevertheless, a gold standard for how people validate human ages.

Even worse, "most birth certificate fraud is committed using genuine documents," and this makes detection so difficult that government officers "fail to detect many fraudulent documents, and the ones they do detect are 'only the tip of the iceberg' . . . their efforts to detect fraudulent birth certificates focus only on obvious alterations, such as erasures, smudges, white-out, misspelled words, offset margins, poor seals, dates that do not match, or photocopies."[24] These problems are so pervasive and long term that they cannot be solved, as "any changes to the [US] birth certificate itself will take essentially a lifetime to become effective."[25] In addition, huge numbers of US migrant communities have no original birth papers through no fault of their own: their governments often never gave them any, or they lost them when fleeing brutal regimes. This problem is not solvable by improving paperwork or checking records, no matter how good US birth registration becomes.

The final nail in the coffin of US paperwork is the fact that these errors are not smoothly spread out at all ages and in all populations. Even vanishingly rare errors build up at the worst possible places in the most nonrandom possible ways, affecting the poorest and the most elderly.[26]

This is why, in 2010, Puerto Rico abandoned birth certificates. The levels of fraud and error had been too high for too long.[27] The wealthiest country on earth just gave up. The Puerto Rican government declared that all birth certificates issued before 2010 were no longer legal documents and restarted the system from scratch.[28]

If the United States cannot deal with millions of identity thefts and faked ages today, using holographic plastic driver's licenses and chip-embedded passports (not to mention the cyber proctology of the National Security Agency), what hope does 1920s Azerbaijan have in doing the same?

I was beginning to be surprised that the problem was not much worse. I soon found examples where it *was* much worse.

In the 1970s and 1980s, Qatar faced a problem: nobody had birth certificates. So, the Qataris simply made up the ages of all their adult citizens. They sent you to the doctor, and the doctor squinted at you and then wrote down your age. Any age. Then, they dashed off a birth certificate. The World Health Organization called the program a success, and other countries followed suit.[29]

China is in between the United States and Qatar, despite an incredibly long history of writing down births and deaths in family records, having their first census around 60 AD, and introducing the "hukou" birth and death registration system in 1951. The hukou system was a two-speed system of registering people that was basically a form maintained by your dad—if your dad was from rural China—and sent to whatever local government existed. When the country was opened up under Deng Xiaoping, the hukou system was not universal, centralized, or standardized: it was, much like the Japanese *Koseki* system, a complete dog's breakfast. It was often only applied inside cities, was applied very differently in the city and the countryside, locked people into inherited roles (and therefore had an incentive for forgery), and was based at the household level of self-reporting. When the one child policy was enacted, identity became even more fungible: people would often go unreported before they became the "children" of random family members. The entire system was not bad—there were certainly worse—but China only changed over to centralized birth certificates in 1996. It was then the most populous country in the world.

India was, and remains, even worse. The most populous country in the world only approached 50 percent birth registration around 2013.[30] Neighboring Bangladesh is similar, just reaching 56 percent of births registered, and Pakistan is much lower, with only 34 percent of births registered in 2018.[31] Two Pakistani prime ministers have confessed to having no idea how old they are, and they are hardly alone. These three countries contain nearly two billion people.

Worldwide, a quarter of children under five do not have a birth certificate today.[32] A third of all births and deaths in Africa remain unregistered, and the continent's most-populated country is ranked as having the second-worst certification system in the world.[33] Some countries simply have nothing. Others struggle with fraud and coverage issues. Some are at war, some have no functioning government, and in some it is possible to simply buy public officials and print yourself some documents on a whim.

Death certificates are not much better. In India, some 30 percent of deaths are still unregistered.[34] Everyone in India born before 2013 is more likely than not to have had no official papers, and almost a third of people just keep on living (on paper) after death.* The opposite case, where a person is registered as dead on paper while they are alive, happened to one of my co-Ig Nobel winners Lal Bihari, who founded a bustling society called the "Association of Dead People."[35]

Basically, we are in a shambles.

Normally, measurement errors like this could be dealt with theoretically. Even the Large Hadron Collider and the James Webb Space Telescope have statistical noise and measurement error. Unfortunately, age is a special case. The particular bear trap the community was falling into, and pretending everything was fine, is this:

1. In most statistical models, you can ignore a high rate of measurement error, provided that the errors are fairly random and symmetrically distributed. You can still extract a meaningful result from noisy data if this condition is even approximately true.
2. In survival processes, errors accumulate nonlinearly and nonrandomly with age, in almost the worst possible way.[36] The assumption in step 1 is destroyed, even if you only have very rare and completely random symmetrically distributed errors. An error rate of 1 in 10,000 is enough to ruin your data because it loads the junk data onto extremely old ages.
3. We have long been incapable of measuring or seeing these errors. Intensive examination of the world's oldest people has repeatedly seen demographers validate fake cases. Often, 100 percent of the "exhaustively

*Other people "die early" on paper—with the effect that British life insurance companies will not pay out if you die in India unless they see your body: people kept going back to the subcontinent and "dying" by purchasing a fake death certificate, in order to enjoy retirement on their life insurance payout.

validated" cases are invisible junk, randomly discovered decades later. Any assumption that a high error rate can be spotted in older ages is therefore directly contradicted by historical fact.

4. Therefore, we have a situation where we *expect* errors to build up at older ages, where we know errors can be common, and where we know (through observation of past failures) that these errors can remain invisible and undetectable for decades.

Staring down the barrel of this logic, I wanted to know how bad the damage was, but I was out of energy.

The answer came unexpectedly. After an uncharacteristic attempt to ignore the mistakes and misconduct of other people, I decided to answer the question that I had been asked more than any other by the public. If I was so smart, the Blue Zones were rubbish, and the supercentenarian regions were all a big spreadsheet error, what were the places with the best old-age survival?

To find out, I analyzed the most carefully curated and exhaustive database in the world: the World Population Prospects Database, a count of virtually every human on earth in every age category, screened and corrected by a team of demographers and public health officials at the UN, and assembled from quality-controlled data sent in by almost every government on earth. Some 236 member states and governments contributed to this database, and the UN records began in 1950. In 2022 it published an update, and I decided to analyze it all.

I did not have to write any code before my impression of firm and careful competence, which I had assigned to the UN, crashed down on my head. Downloading the data and sorting them revealed the UN estimates for longevity at age one hundred and above. What countries, according to the UN, currently have the best life expectancy at age one hundred?

New Caledonia, Puerto Rico, and Thailand.

Puerto Rico, where they canceled the birth certificates because of all the fraud. I felt like all those dogs who get invited for a "walk" and come back spayed. I had learned nothing.

I figured there could be a lot of noise. Please, let it be noise. After all, there were not many people aged one hundred and above, so perhaps it was a random sampling effect. I decided to clean up the estimates, so I measured rates of survival from age eighty to one hundred and older in every country on earth. I called this the "centenarian attainment rate" for simplicity because it measured the odds of an eighty-year-old reaching age one hundred and above.

Who has the best centenarian attainment rate in the world?

It was Monaco. Long-lived, disgustingly wealthy Monaco.* I breathed a split-second sigh of relief. Then I saw the rest of it. In second place was Thailand: a happy middle-ranking country for health and 54th globally for life expectancy at birth, before Guadeloupe (38th for life expectancy) and Kenya, which was the 213th country in the world for life expectancy at birth out of a possible 236.[37]

Life expectancy at birth is probably the most centrally important metric of human health. By this metric, Syria, in the middle of an ISIS insurgency and a civil war, was ranked 132nd in the world. Kenya was ranked 213th yet ranked fourth for old-age survival at the same time.

Malawi routinely made it into the top ten for centenarian attainment—it has a lower GDP than North Korea and ranks 202nd in the world for life expectancy. It had managed this trick almost every year since records began. So had Western Sahara: a Wild West stretch of sand that *does* have the longest land-mined wall fortification in the world but, awkwardly, *does not* have a government. In the 1980s it was the best place in the world for eighty-plus-year-olds, according to the UN data. The rebel POLISARIO Front was then fighting in a ferocious guerrilla war against Morocco, so I guess I had found another secret to long life: sand and landmines.

The results were not sampling errors. They were not artifacts. They happened year after year, cohort after cohort. Despite massive growth in the number of 100-plus-year-olds worldwide, the data remained absurd from start to finish, 1950 to 2021. Rwanda made the top ten for centenarian attainment in the middle of the AIDS crisis.[38] Dictatorship-mad Turkmenistan entered the top ten in the '90s. Mexico routinely demolished the United States in the late-life survival rankings.[39]

---

*Just to add insult on top, Monaco only seems to be at the top of the ranking because it has high migrant inflows at old ages, which uniquely bias the estimates for this country. People move to Monaco to avoid paying inheritance tax—which can be ~50 percent in France but is 0 percent in Monaco—resulting in a lot of eighty-plus people moving there. No such patterns explain Malawi, Thailand, or Western Sahara.

It was not just the top ten rankings either. The whole thing was a basket case. If you looked at countries with short lifespans, they were everywhere: from the best to the worst places for centenarian attainment rate. If you looked at countries with long lifespans, the rankings were a mess. Sweden was ranked 86th in the world for late-life survival. Super-rich Liechtenstein was 106th. Uruguay was, somehow, 6th for centenarian attainment rate.

These were, and are, the UN's and global health systems' best guess.[40] If you asked the UN where the real Blue Zones were, its data consistently said Thailand, Puerto Rico, Guadeloupe, and Malawi (figure 7.1)—four of the most random countries imaginable. The only thing these data resemble is an error process.[41] The horrible conclusion is that our survival data were unusable beyond age eighty, at all scales and in all time periods and in all countries.

I began to despair. Had I done something wrong? Was I mad? If I were mad, why was I employed at the University of Oxford? In some cases, my analysis was so simple that it could be reproduced by carrying out the task: "download the data and click sort in Excel." I could hardly conclude I had made a complicated and idiosyncratic mistake. I asked my colleagues to provide me with an answer, any answer. To show me the error in my code. To reproduce what I had done. I was surrounded by professors of medicine, demography, and data science. Some were friendly, some were brilliant, and some would jump at any opportunity to throw me under the closest available bus. They were all silent.

What special insanity was this?

Finally, a few people decided to save me. They ran the code. The math checked out, the theory was solid, and the results were reproducible. A couple of data teams at major newspapers did the same. Some of these experts had risen to prominence by debunking pseudoscience during the pandemic, and I felt I could trust them to be brutally honest. Everyone reached the same conclusion.

The answer to the public's question "what are the best places to reach age one hundred?" was a disappointing "nobody knows." Beyond age eighty, none of our data are any good.

It was a disaster. Our old-age mortality records are junk. They always have been.

Age misreporting had highly nonrandom effects in the UN data, turning our basic picture of human health upside down. The collective governments of

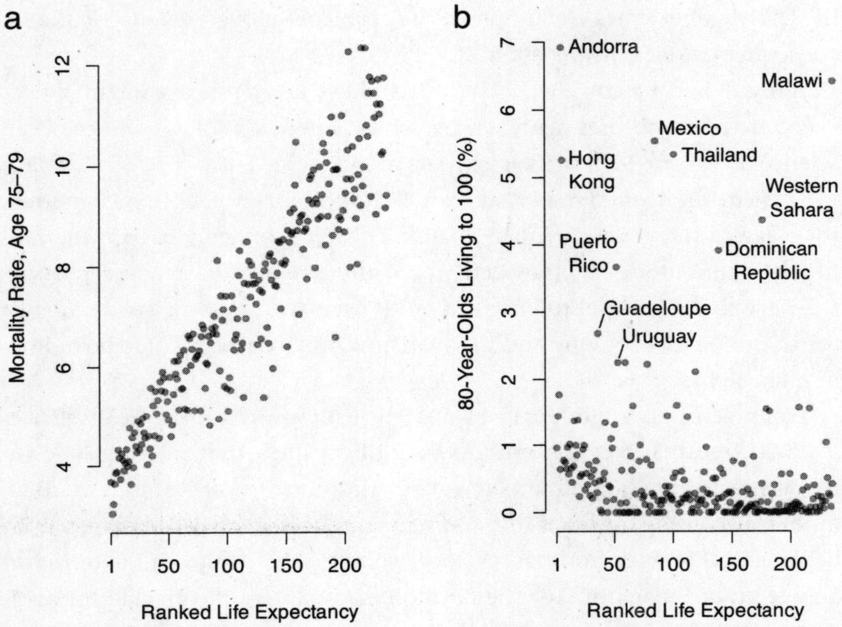

Figure 7.1

The most likely places to reach 100 in the world, for the year 2000. Analysis of UN data reveals the expected relationship between life expectancy at birth, ranking from best (#1) to worst (#236), and mortality rates up to age 75–79 (shown here for the year 2000 in panel a). As you may expect, worse life expectancy accurately predicts higher mortality rates at these ages. Using a statistical measure, we can see that 87 percent of the variation in mortality rates at age 75–79 (panel a) is accounted for by life expectancy at birth (adjusted $R^2 = 0.87$). Then things fall apart. If we measure the percentage of people surviving from ages 80 to 100 in the same data (panel b), the best places in the world for surviving to age 100 are a statistically random pile of countries. I have labeled the top ten for the year 2000—the UN Blue Zones—and they include a scattergun sample of countries. Worse, nothing in old-age survival is predicted by earlier mortality patterns, with just 4 percent of the variation in reaching age 100 predicted by life expectancy (adjusted $R^2 = 0.04$). The message is that beyond age 80, our mortality estimates are basically junk. Source: Author.

the UN member states could not see the problem for over seventy years and consequently did nothing about it.

Sadly, I had to consider what this hidden error process might do to everyone *else* who uses age data, like those in epidemiology and medical science. Everyone I knew took age data at face value.

Even in the United States, at least 5–10 percent of people misreported their age to the government in 2020.[42] This level of age misreporting has highly nonrandom effects. What would these people do when reporting their age to a scientific study or survey? If even a 0.1 percent error rate distorted our picture of aging and survival, how much damage could be hiding in plain sight?

Using a search engine called PubMed, which indexes medical science research, returns over 3.4 million scientific papers that mention age in the abstract. This is approximately $2.2 trillion worth of research in 2024 money. Becoming more specific and searching "age related" hits over a million medical science findings, as does "cohort," the demographic term for an age group. Searching for the far more specific term "corrected for age" gives 110,000 papers, narrowing again to the exact phrase "stratified by cohort" turns up 45,000 scientific papers. Even the extremely specific "over age 80" returned a staggering 409,318 hits for scientific papers.

Only a tiny fraction of these had to be affected by age errors to generate a systemic disaster. Yet it was clear that nobody was listening.

The realization was crushing.

The billion-dollar industry of Blue Zones was not important. Even the fun circular logic of the demographers was a side note. I had let myself be distracted by guff, from something utterly shocking in scale, and by the time I tried to raise the alarm, I was too professionally isolated to have any effect. I had math, open data, and open code on my side. However, I was threatening an industry, and nobody was going to speak out for me in public. The seventeen continuous years of racism propagated by Robert Young had been met with silence. I had very little chance in convincing people of something *far more* uncomfortable.

The more important stuff, the large-scale patterns of fraud and top-tier cases of misconduct, were never going to be mentioned in polite society.

There were a few bright lights of support—Professor Charlie Rahal and Professor Sarah Harper especially—but I retreated. I published papers as a distraction, alone or in other fields, and tried to deal with my severe depression.[43] Something was profoundly wrong in aging research, but I could not gain any traction. I was alone, and I was not going to get less alone. I had to do something or let it all go. Eventually, I did let it go.

Internally, I quit demography. The toxic fiefdoms had won.

I still wanted to do something good with my life, so I retreated to the quiet splendor of the Oxford libraries and started other projects.[44] I began applying for new jobs as a plant scientist, a geneticist, or a medical scientist. After a close-ish call with an assistant professorship in plant science, I landed a position at University College London as a senior research fellow in medical science. I had kept a desk at the Oxford Institute of Population Aging, but I left demography in despair.

Three people attended the farewell.

I won the Ig Nobel Prize a few days later.

8 Peach Groves and Virgin-Boats

What we call "age validation" is not validation at all. It is a check to see if age records match: a check for consistency. As if looking at an old bathroom scale with a worn-out spring, aging science mistook consistency for accuracy, and bad measurements are now everywhere.

One recent example is a wildly successful pair of papers on human cardiovascular aging, published in two elite scientific journals, the *Lancet* and the *Proceedings of the National Academy of Sciences*, that studied a classically exoticized culture: a tribe in the Amazon.[1]

These studies examined the heart health of the Tsimane tribe who—like the Vilcabamba residents who faked and exaggerated their ages for Harvard's Professor Leaf—come from remote regions of South America where documents are (ahem) rare.[2] Using detailed scans of cardiac function, the studies found that the Tsimane had remarkably good heart health for their age. According to the scientists, "a 75-year-old Tsimane's arteries are more like a 50-year-old American's arteries."[3] Unfortunately, this assumes that the scientists have an accurate idea of the *real* age of the Tsimane people involved.

So what if, and stick with me here, the recorded ages of remote tribespeople in the Amazon are not accurate? What if the Tsimane residents have remarkably good heart health "for their age" because "their age" is being exaggerated by people with severely limited literacy and numeracy, who live in a culture where being a "village elder" gives you huge status and prestige? The entire study would collapse. Nobody gets a string of massive National Institutes of Health grants, nobody gets to do fieldwork in Bolivia, and nobody gets to overly exoticize a remote tribe with a worse average life expectancy than every single country on earth.[4]

In the words of the researchers measuring Tsimane ages, "many people do not know their ages" and were included in the study anyway. Every year, the researchers ran a survey of the Tsimane people.[5] Over time, roughly half of the Tsimane made up impossible ages, routinely inflating their age by decades. Incredibly, the researchers did not remove anyone but instead took the median of the impossible-and-self-reported guesses as the real age.[6] If a Tsimane tribeswoman was 40 in 2009, then 55 in 2010, and then 62 in 2011, they assumed she was 55. If this incredibly dubious method failed as well, because someone was 128 years old or aged in reverse, they still didn't remove any cases and instead came up with increasingly strange and unvalidated age estimates instead.[7]

By 2017, the Tsimane dataset had spawned "109 peer-reviewed publications in 42 scientific journals."[8] Most of the papers used age as a variable or had age data as a core measure, including the *Lancet* paper.[9] The likelihood is that most of these results are tainted and unusable because the scientists included ages that they knew were invented.

This inability to question official records is a common cultural blind spot, in which rich folks from rich countries cannot conceive that a set of official records might be inaccurate. Instead, some of these researchers have propagated a narrative in which, apparently, the wisdom of the rainforest makes you age more slowly.[10] To them, this cardiovascular near-miracle is somehow more plausible than a person in the rainforest exaggerating their age, or the Bolivian government keeping bad records for an ethnic minority.

There are tens of thousands of other studies in which ages are taken at face value. Many have a similarly poor chance of withstanding scrutiny if those ages turn out to have systemic errors. We have no way of discovering such errors, but you hopefully know that by now. But there is much more to this story than bad data.

Everyone seems to be ignoring a much tougher question: what kind of research culture can fail to see problems of this magnitude for decades on end?

The problem with the Tsimane papers is not just a technical blind spot. You do not need to understand a mathematical process or an arcane statistical model to understand that including invented, ever-changing age data is a terrible idea. What kind of research culture allows you to make up a key variable like age, without anyone involved raising an alarm, and publish a hundred papers off the back of it?

This is a huge cultural failing, where people "overlooked" bad data to perpetuate a myth.

The Blue Zones research spanned many fields: for over twenty years, they made refutable claims in genetics (e.g., incest increases survival), medicine (e.g., heavy drinking, smoking, and goiters[11] are linked to health benefits), nutrition (e.g., Okinawans eat lots of beans/vegetables/sweet potato), geography (e.g., the Sardinian and Costa Rican Blue Zones move around when it is convenient for the researchers), mental health (e.g., Okinawans are happy), and economics (e.g., poverty and hardship is a net positive for survival) without being questioned or called out.

Nobody, in any of these fields, appears to have stood up and questioned why a popular concept was at odds with the evidence in their own scientific field. The entire UN apparatus for counting people and measuring human health estimated that the best places to be elderly was in communist dictatorships, government-free states, countries at war, and postcolonial islands with systemic poverty. Again, nobody said a word for decades. It seems nobody even noticed.

Forget the demographers. Forget the longevity research gurus. Where are all the doctors, public health officials, or epidemiologists calling out these ideas? Why were the Blue Zones data issues, or the issues with the global collection of data held by the UN, or even the niche but obvious problems with the Tsimane "healthy tribesmen" studies, not discovered by other people in other ways? These studies are not an outcome of individual greed or stupidity, though these are important factors. They are emblematic outcomes of a systemic failure.

The ingredients are simple enough. A substantial part of humanity does not want to die. With a slightly weaker preference, another large fraction of humanity does not want to go jogging, stop smoking, or cease drinking either—all the things we know extend our lives already. We all want to escape death, but most of us don't want to get sweaty in the process.

Another sizable percentage of humanity, the highly motivated, put in the effort to delay death through exercise and sobriety. Unfortunately, they rarely stop there and almost always suspect that they need to do something *extra*. If you run a weekend marathon for your health, the odds are you will go even farther, and spend even more, to delay death.

This creates a permanent opportunity for myths and marketing.

Our most ancient myth of all is Gilgamesh seeking to take a flower of immortality from the deathless Utnapishtim: a person living a simple life far away, with a herby secret to immortality. Yet it is two Chinese scholars, Sima Qian and the Daoist scholar Dongfang Shuo, who beautifully describe the first great longevity grift.

The first Chinese emperor, Qin Shi Huangdi, obsessively sought the "drug of deathlessness." Like many of the megarich, and many Chinese emperors after him,* the first emperor was fooled by the unscrupulous sales pitches of "travelers, magicians, and alchemists" selling cures for death. He drank the "cure" of mercury sulfide dissolved in wine, and searched for secrets of immortality by executive decree,† but the best sales pitch to the emperor came from Xu Fu. In a bold gamble, given the emperor's tendency to bury scholars and alchemists alive, Xu Fu asked for "3,000 virgin boys and 3,000 virgin girls," money, and a fleet of sixty ships to go find the cure for immortality on the mythical Mount Penglai, out upon the Eastern Sea. Why he needed virgins was never exactly explained, but I'm sure he had something in mind.

Xu Fu left and didn't find the cure, but not content with pleasure cruise number one, he came back for a second round or seed funding. Xu Fu claimed to need more resources because, in one of the best excuses ever made, he had been "blocked by a giant sea monster." This time, Xu Fu took the virgins, ships, and extra money; sailed over the horizon; and wisely never came back.‡

I think of Xu Fu whenever I underestimate the audacity of a good longevity scam.

---

*At least fourteen emperors of China have been hooked on the "immortality" sales pitch since Qin Shi Huangdi, many of whom died as a result. According to court scholar Zhang Tingyu, the last Chinese emperor killed by a crap off-brand immortality elixir was Emperor Yongzheng in 1735.

†Archaeologists retrieved one of these decrees from a well in Hunan province in 2002, along with some awkward replies from the officials tasked with making the emperor immortal. A place called Duxiang came up empty for immortality juice, and Shandong province just sent him a bunch of random herbs.

‡According to Japanese legend, Xu Fu sailed his fleet into the Ariake Sea behind Nagasaki, floated a cup on the water, and followed it ashore to land in what is now Saga city. He climbed Mount Kinryu and found the secret to immortality. Apparently, this time it was *Asarum nipponicum*, or Japanese wild ginger: a pungent and beautifully patterned plant that produces the kidney-destroying aristocholic acid. With his questionable salad nibbled, Xu Fu settled down to be a part of the local nobility instead of returning to the emperor, with no mention of the fate of the virgins.

Not much has changed in the last 2,230 years. We still buy the herbs. We still buy that the secret lies far away in an exotic place. We still give our money to Xu Fu and watch him sail away.

The alternative is jogging, so what else are we to do?

Just-so stories seem to survive in aging research because a sizable segment of society, including scientists, want to believe in such myths. If the alternative is exercise, sobriety, and the unflinching acceptance of your own inevitable death, a considerable market share of humanity will buy an "antiaging" fairy tale no matter how silly it becomes.

Nobody wants to see the evidence because it would shatter the myth. Yet it is a disservice to medicine and to ourselves to ignore the bad data and logic. Myth-smashing also gives us new mental tools to defend ourselves against any new forms of bad logic and herb sales.

Unfortunately, for tedious legal reasons, I cannot just decry a list of obviously fake solutions by name, but I can show you a recent history of aging claims and let you decide which ones are bunk. Let's learn to spot any new Xu Fus before they sail away with the virgins.

9 Mousetraps

From birth until death, aging causes the risk of death to double about every seven or eight years of age in humans.[1]* This process is called the mortality rate doubling time (MRDT), and it goes forward like clockwork.[2] Imagine you have an MRDT of eight years. Get older, from fifty to fifty-eight? The odds of death double. Go from fifty-eight to sixty-six? They double again. From sixty-six to seventy-four, another doubling; seventy-four to eighty, another doubling—so that your day-to-day odds of dying at age eighty-eight are sixteen times (1,600 percent) higher than when you were fifty. The odds of dying keep going up like this, from about age thirty, until you die.

This mortality rate doubling is caused by aging and is expressed as a near-constant rise in mortality risks from diverse causes of death—like heart attacks, cancers, stroke, Alzheimer's disease—all stacked together. The risk of death from virtually everything goes up with age, and when you add these risks up, the result is a near-exact and clockwork doubling of the risk of death every few years.

Aging causes this rise in mortality risk. And so if a scientist, doctor, quack, or mystic claims that an intervention prevents or slows or reverses aging, they will have a treatment for *every single one of the diseases caused by aging*.

This is a useful lens to filter claims.

Let's look at a typical claim by a (hypothetical) professor who said he could lower his biological age by about ten years. Imagine what would happen if this were true.

---

*Other mammals also hit a point where mortality rates double with a linear increase in age: mice, dogs, rats, elephants, and monkeys all have an MRDT but have a different value to the human MRDT. In mice it is about 120–160 days, though it depends on the type of mouse.

Mortality rates double with age because of the aging process. Therefore, if his intervention reduced "biological age" by ten years, his magic pill would have also cut the risk of death by more than half. Not just for one disease but for basically every disease. Think of how utterly insanely profitable this pill would be to a pharmaceutical company.

The hypothetical professor could run a clinical trial, showing he had halved the risk of death from the hundreds of types of cancer caused by aging. He could do the same to show he had halved the risk of heart disease. Stroke. Chronic obstructive pulmonary disease. Every common disease we now suffer, and which gets worse with age, would be treatable. It would not just be the fatal diseases either: everything caused by aging, from erectile dysfunction to sleep apnea to wrinkles, would respond to the pill.[3] After his clinical trials were complete, his "ten-years-younger" drug would be worth hundreds of billions of dollars, and he would win a Nobel Prize or three for basically solving medicine. The Food and Drug Administration (FDA) would approve the drug, and every sensible doctor in the world would prescribe it. He would become insanely rich, insanely famous, and be lauded as a hero.

So, how do we explain the fact that our little professor has not made his billions or, at the very least, has not completed a very well-funded clinical trial with stellar results? If these magic pills had any basis in fact, pharmaceutical companies should be smashing down the door of aging researchers to get a slice of the antiaging riches. People would pay anything, literally anything, to not age.

Instead, megacorporations like GlaxoSmithKline, Merck, and Pfizer no longer fund antiaging studies, "old finance" money avoids antiaging science altogether,* and our little professor is selling his "secret" as a supplement on social media channels and lifestyle podcasts.

Because our little professor is lying or, if we are very generous, fooling himself.

To avoid fooling ourselves, let's explore how a bad understanding of statistics can make bad aging research.

---

*Antiaging funding is largely the domain of confused tech billionaires or cryptocurrency types with unhinged social media feeds who flog pump-and-dump investment start-ups. Established finance avoids this stuff like the plague.

Imagine I have a mouse. Let's say it exists in one of two clearly defined states to avoid any irrelevant metaphysical arguments: the mouse is alive or dead. At some point the mouse goes from one state (living) to the other (dead).

You can imagine two components for how fast the mouse arrives at a "dead" state by picturing a car accelerating at a brick wall. Two things will decide the time taken to hit the wall: speed (the mortality rate, an instantaneous measure) and the acceleration rate (the aging rate, which is *a rate of change* in the mortality rate).

The core problem of aging statistics is that in medical science, we rely on statistical models that measure the time taken for a living mouse to become dead: their survival time.

Yet if a car hits a wall faster than another, we do not know if it was because the car had a higher (possibly constant) speed, a higher acceleration rate, or both. Likewise, if a mouse dies faster, we do not know if it was because of a higher (possibly constant) mortality rate, a higher aging rate, or both. All we know, and all we normally measure, is the time that has elapsed before death. Comparing the survival rate of two mice or two groups of mice is therefore a very shaky guide to differences in aging rate.

Let's unpack this a bit more because it is key to understanding the entire problem.

For example, let's imagine we applied an intervention, found that it made a group of mice live longer (statistically), and concluded that it slowed down the aging rate. Does this logic hold together?

It may sound good, but the answer is no.

To give us a concrete idea of where the logic goes downhill, let's choose a popular idea with a clear experimental design,* published in a top-tier journal.[4] The paper's finding was that two popular interventions, caloric restriction and intermittent fasting, are linked to higher survival rates in mice. Caloric restriction is eating fewer calories while having enough micronutrients and vitamins, whereas intermittent fasting is when you regularly don't eat for short periods, say fourteen hours.[5]

All of which sounds great. But let's do something funny. Let's play copy-paste and swap "caloric restriction" and "intermittent fasting" with some

*I am picking on this paper because it later does the right thing, almost, by measuring the MRDT, with a too-small sample size.

extrinsic risks of death for mice, ones that have nothing to do with aging. Like, say, "fewer snakes" and "more mongooses," respectively. The introduction of the paper now reads:

> ~~*Caloric restriction*~~ [Fewer snakes] *extends healthy lifespan in multiple species.* ~~*Intermittent fasting*~~ [More mongooses], *an alternative form of* [fewer snakes], *is potentially more sustainable in humans, but its effectiveness remains largely unexplored. . . . Here we performed an extensive assessment of graded levels of* [fewer snakes] *(20% and 40%* [fewer snakes]*) and* [more mongooses] *(1 and 2* [mongooses] *per week) on the health and survival of 960 genetically diverse female mice. We show that* [fewer snakes] *and* [more mongooses] *both resulted in lifespan extension in proportion to the degree of* [snakes].[6]

If you ran this experiment with the appropriate snakes, mongooses, and mice, would you get similar results? Yup. Sure. Would the hypothesis generate similar survival curves and statistical tests? Of course. Does this mean snakes cause aging? No.

You cannot conclude that an intervention caused a mouse to "age slower" because it lived longer.

You can play this game with tons of papers, thousands even. When there is only one intervention, I prefer to use copy-paste on "mousetraps" instead, as they exist at a low density in the environment and are easier to imagine as a stationary random threat.

Let's play again. This is the introduction to a now-famous paper on rapamycin, a supposed antiaging intervention of incredible popularity, being tested in 2009 by the National Institute of Aging under a special framework aimed at finding lifespan extension methods.

> We report here that ~~rapamycin~~ [fewer mousetraps], an inhibitor of ~~the mTOR pathway~~ [mousetraps], extends median and maximal lifespan of both male and female mice when ~~fed~~ [introduced] beginning at 600 days of age. Based on age at 90% mortality, ~~rapamycin~~ [fewer mousetraps] led to an increase [in survival] of 14% for females and 9% for males. The effect was seen at three independent test sites in genetically heterogeneous mice, chosen to avoid genotype-specific effects on disease susceptibility. Disease patterns of ~~rapamycin-treated~~ [mousetrap-free] mice did not differ from those of control mice. In a separate study, ~~rapamycin fed to mice~~ [fewer mousetraps] beginning at 270 days of age also increased survival in both males and females, based on an interim analysis conducted near the median survival point. . . . These findings have implications for further development of interventions targeting ~~mTOR~~ [mousetraps] for the treatment and prevention of age-related diseases.[7]

Again, we see that if we varied the density of mousetraps instead of the original intervention, we could produce similar results. The error here is the following:

1. Faster aging is when things *get worse faster with age.*
2. If we keep the aging rate constant (things always get worse faster with age at the same rate), we can still make things *worse overall* (or better overall) without changing the aging rate. For example, we could add mousetraps to the environment.
3. Therefore, higher mortality rates and shorter survival can be caused by things *being worse overall* or by things *getting worse faster* with age.

You cannot tell which situation is happening just by using the normal survival tests used in aging research: by measuring average longevity* or by measuring the age of the last survivor in a population. These are instantaneous or period-averaged measures of survival and cannot capture aging rates.

The National Institute of Aging's Intervention Testing Program (ITP) is perhaps the clearest, but by no means the only, demonstration of this shortfall in statistical clarity.[8] The program has been running for two decades at immense expense and claims to be "an aging interventions testing program" developed by the premier aging science body in the United States.[9] Yet the program does not measure how fast things get worse with age.

The ITP compares groups of mice for how long they live, sure, but does not measure how fast their mortality rates accelerate with age. It is mousetrap research.

To clarify the problem, let's take the ITP's very first significant "antiaging" result, achieved by giving two drugs (aspirin and something called NDGA, or nordihydroguaiaretic acid) to mice, and reanalyze the data. The ITP trial marked aspirin as "non-steroidal anti-inflammatory† drug, anti-aging,"

---

*We have a surprisingly poor idea of what environmental hazard "mousetraps" are killing our lab mice. If, for example, a common cause of death in lab mice were autoimmune diseases, then we could suggest that rapamycin may be raising survival because it is an immunosuppressant and not due to antiaging. Yet we are largely in the dark.

†It is not possible to skip past Derek Lowe's great op-ed that "mouse models of inflammation are basically worthless" because "mice are not humans" and share no inflammation responses at all—published in *Science* in 2013 and still ignored today.

called NDGA "anti-oxidant, anti-aging," and used something called the log-rank test to measure differences in survival rates.[10] Yet this test cannot measure whether two populations are aging at different rates.

I can,* so let's compare the rate that mortality gets worse with age in the control group to the same rate in the aspirin- and NDGA-taking groups. First, we find that the sample sizes are too small to get a reliable result. There are only 621 dead mice observed in the control group, 308 in the NDGA group, and 308 in the aspirin group. This seems like a lot to a biologist, and you can certainly see which group has worse survival times *overall*, but to a demographer it is not enough to get a robust estimate of how fast things get worse with age. The rate at which mortality rates increase with age is not significantly different between the groups, which is unsurprising given that the statistical design is extremely underpowered.

If we proceed with the analysis anyway, what do we find? If we just ignore the too-small population and perform a calculation regardless, we find that the mortality rate doubles every 131.9 days for the NDGA mice, every 140.3 days for the aspirin-munching mice, and every 140.6 days for the control mice. That is, the two groups taking the "antiaging" drugs had survival rates that decayed *faster* with age compared to the control mice. Survival was better for the drug-taking mice overall, but it got worse faster with age. The drug mice age faster, from a slightly healthier baseline. The ITP study never seemed to have any idea.[11]

Unfortunately, even this contradictory result is not statistically significant: as the numbers are just too small, it could be random chance. So, our antiaging National Institute of Aging intervention program has revealed nothing about aging, and our only indication of any real effect (if one exists) is the opposite of the claimed result. Regrettably, the ITP has been doing this kind of ambiguous mousetrap research for twenty years.

To understand just how big of a problem this is, we need a metaphor. Think of mortality rates like a batting average in baseball or an equivalent measure of success in any other sport—sprint times in running, driving distance off the tee in golf, or whatever you like. They are an instantaneous measure of success.

Now imagine, for example, you have a performance-enhancing drug that makes your batting average better for a certain amount of time. Would this

*I have added the code to GitHub and can send it to anyone who requests it.

automatically mean your batting average declined more *slowly*? Of course not. The two things are unrelated. One is a measure of how good you are at one instant, and the other is a measure of how fast you change with age.

Measuring one does not help you understand the other. Yet in aging science, people routinely confuse time-averaged survival metrics (the "batting average" of longevity) with aging rates (the rate stuff gets worse) and imagine that one tells us about the other.

As a result, for decades, scientific studies on aging have behaved as if differences in overall mortality rates—measured by statistics with fancy names like the log-rank test or the Kaplan-Meier survival curve or Cox proportional hazards test—measure aging rates. They don't measure anything of the sort.

A small number of people have noticed. A recent evaluation of mouse models of aging—inbred lines of mice that are supposed to age faster or slower than their otherwise identical cousins—revealed that at least half were mousy frauds in 2018. Comparing "fast-aging" and "slow-aging" groups revealed no differences at all in aging rates: "only 15% of the genetic manipulations analyzed significantly affected the demographic rate of aging as predicted . . . manipulations affecting longevity tend to impact on aging-independent mortality."[12] Most of the differences in mouse "aging" models were due to genetic causes of death that—like a mousetrap—killed mice without affecting the rate of aging.

Given the huge amount of research based on these mice, and given that it takes at least five years and enormous cost to even generate a new mouse line, this is a disturbing result. "Mousetrap" research has had another effect, however, beyond wasted money and empirical confusion. Statistical misunderstandings have allowed people to take improvements in overall mouse survival (or other organisms) in the lab and promote these findings as if they were remarkable breakthroughs in aging research.

10 Venture-Capital Vampires of California

You cannot explain the existence of modern immortality elixirs if you focus on the science alone because, sadly, the science is usually peripheral. The specific chemical composition of the elixirs hardly matters and is often interchangeable. What really determines the success of sales pitch mousetrap science is the integrity and personality of the people involved. Therefore, to chart the modern Xu Fus trying to sell immortality, it is necessary to focus on personalities. This is generally frowned upon in science writing, but to avoid it would ignore the central role that personality plays in aging and longevity research.

Let us begin with one of the most central and hugely successful personalities and risk looking a little unscientific in the process.

David Sinclair began his career in a lab run by Professor Guarente, a man whose memoir is called *Ageless Quest*, who would make two of his students watch yeast cells continuously in rotating twelve-hour shifts, who got in the lab at 8:30 and left at 12:30 in the morning, and who shouted relentlessly at his underlings. As a Harvard postdoctoral fellow, Sinclair fit right in. He published a high-level blockbuster paper—which he reassures everyone was definitely not stolen from a postdoc in the same lab who had "proposed similar studies"—and got a permanent post at Harvard from Professor Guarente's "glowing recommendation."[1]

Sinclair soon began fighting with his mentor Professor Guarente, who had just raised $40 million through a company called Elixir to target a "longevity" gene he discovered. The premise was that one gene somehow controlled aging,* and Guarente had started a company to exploit the hype and make us all live longer.[2]

---

*Upregulation of this gene has since been shown to make yeast (and fruit flies) live shorter, not longer, after decades of hype stating the opposite.

The reason for the fight was that Sinclair claimed he had discovered something huge. It was, he said "as close to a miraculous molecule as you can find. . . . One hundred years from now, people will maybe be taking these molecules on a daily basis to prevent heart disease, stroke, and cancer."[3] The molecule was called resveratrol and could be found in moderately high quantities in grape skins and wine sediments. It supposedly targeted the same "immortality" gene as Guarente's "Elixir."* Guarente would eventually leave when Elixir began to "just import stuff . . . marginally related to aging, and call that an anti-aging company" before the company dissolved, immortality-cure free, in 2018.[4]

Yet Sinclair's resveratrol company would fare much better. After all, it was an antiaging drug extracted from red wine—a miracle molecule that sells itself, especially if you go flogging the stuff to Californian wine moms on *Oprah*.

To start selling resveratrol, Sinclair got involved with Longevinex: a mail-order supplements company run out of a Las Vegas strip mall, opposite a self-storage place and behind the parking lot of the Jackpot Bar and Grill, run by someone named Bill Sardi. Sardi is also a strange bedfellow for a Harvard academic. When he's not posting creatine, Sardi publishes books like *The New Truth About Vitamins & Minerals* and *The Power of Healing the Power of God* [sic], and appeared on a documentary claiming the COVID-19 pandemic was all a hoax, so take Sardi with a mail-order, ultra-large, bulk-purchase dose of skepticism. Yet according to Sardi, "in 2004 a Harvard professor announced to the news media that he had developed a nutriceutical version of resveratrol. . . . Harvard authorities intercepted this professor's e-mails and phone calls and directly informed him, if he chose to work with a dietary supplement company, that he would never become tenured."[5] In a huge coincidence, Sinclair worked for Longevinex as a paid consultant before he cut ties with the company and claimed he "received no compensation from the Longevinex company."[6] Sinclair then got tenure at Harvard in 2008.

Sinclair promptly founded his own company, called Sirtris, to rival Guarente and sell resveratrol. He ended up owning just 1 percent of Sirtris—the venture capitalists got the rest—but they raised $105 million in start-up

* It was subsequently shown to do nothing of the sort.

capital and rapidly inflated the share price through hype and TV appearances when they floated the company publicly. Sinclair even promoted Sirtris and resveratrol on the *Oprah Winfrey Show* with the triple-award winner for misinformation in science, Dr. Mehmet Oz. The response was so huge that Sinclair and Bill Sardi teamed up to try and stop the flood of people cashing in on their marketing by (ironically) selling it as a cheap supplement on the internet.

Yet Sinclair had bigger ideas for his marketing than outcompeting supplement stores. In 2008 the British mega-pharmaceutical giant GlaxoSmithKline (GSK) was convinced to buy Sirtris. It paid $720 million for it, 185 percent above its stock market valuation, and an incredible $1.1 billion in today's (2024) money. Sinclair made at least $8 million for himself.

Something irregular then started to unfold. Despite the wild success, Professor Sinclair then gave a 2009 TED talk in which he claimed that he, Oprah, and Dr. Oz "don't endorse this product [resveratrol], we don't endorse any products."[7] This was an odd claim for the guy who just promoted and sold the company synonymous with resveratrol, and who was later revealed to be earning $297,000 a year in consulting fees from GSK's resveratrol testing division.[8]

A rapid reassessment of the drug was about to take place for the buyers at GSK too. It was soon discovered that resveratrol was not absorbed into the body at meaningful rates, and taking more of it did not improve matters because "extensive metabolism in the intestine and liver results in an oral bioavailability considerably less than 1 percent. Dose escalation and repeated dose administration of resveratrol does not appear to alter this significantly."[9] The miracle compound was digested and "removed," or if it made it into the blood, rapidly destroyed in the liver and peed right back out. It was absorbed at such low rates and excreted with such speed that people justifiably asked if the compound had any biologically meaningful effect at all, and consuming more of the stuff did not lead to absorbing a higher dose—it just led to faster excretion.

This did generate a very reproducible effect during clinical trials. A scientist noted that "resveratrol has the potential of mediating a response analogous to a biological tsunami," which is, ironically, what it does: it gives you horrible nausea and diarrhea.[10] This miraculous outcome seems to be the main biochemical mode of action for the drug[11] and is reproduced in virtually every clinical trial.* Even a paper unironically titled "Resveratrol: A Miraculous Natural Compound for Diseases Treatment" merely generated the strongly reproducible effect of diarrhea.[12]

Resveratrol-induced diarrhea led to tragic outcomes when GSK administered high doses of the expensive new drug to myeloma patients in a clinical trial. Over 80 percent got nausea and diarrhea, and half reported serious adverse events, including "renal failure acute" in three of the twenty-four participants.[13] Two people died and doctors ended the clinical trial early for safety reasons to prevent more deaths.[14] In the aftermath, GSK noted that the miracle drug "was not well tolerated, and side effects of nausea/vomiting/diarrhea may have indirectly led to dehydration, which exacerbated the development of the acute [kidney] failure" in patients.[15] That's grim to hear from an industry spokesperson, and the word "indirectly" looks like it was desperately inserted by a lawyer. The company also noted the drug had "minimal efficacy," which is a bit obvious.[16]

After the clinical trial of resveratrol was halted for kidney destruction and dead people, Sinclair gave the molecule a temporary downgrade from "miraculous" when he said that it "was and remains an excellent proof-of-concept molecule."[17] The drug would fortunately recover its miracle status later.[18]

As for GSK, it would soon realize what it had really bought. By 2011, the scientists at GSK testing resveratrol reached the point of despondency where they told *The New York Times* "what is the actual use for humans still needs to be discovered"—a quote that sounds like God making an entry in a lab book.[19] It's also quite a thing to say about your own product. The company cut their considerable losses and got out. Sirtris, the company it had bought at enormous expense, was eliminated after just five years.

The hype machine carried on undaunted. Sinclair was officially a "big name," and unlike many other antiaging drugs that are basically unfundable, resveratrol reached clinical trials over 130 times.[20] So far, resveratrol and its derivatives do not have any clear clinical use—the market for diarrhea drugs is too plugged up—and have only generated a clinical warning

---

*Resveratrol causes diarrhea and nausea in humans. If it did the same in resveratrol-dosed mice, they would eat less and/or absorb a lower number of calories—a "caloric restriction"-like outcome. Unfortunately, this hypothesis may be unlikely, not because it couldn't happen but because, unlike many resveratrol papers, the effect would be easily reproducible.

not to take the stuff when you're on anticoagulants. As a result, the real-world outcome of well over ten thousand scientific papers on resveratrol has been that "numerous commercial products are marketed to the general public with suggestions of life extension as well as a multitude of other health benefits. Nonetheless, at the present time, there is no consensus regarding the usage of resveratrol based on scientific evidence."[21]

By 2017, the conclusion was that "despite an abundance of laboratory and animal research, there is little clinical evidence that resveratrol is an effective therapeutic in humans."[22] For context, I could probably run 130 clinical trials on the capacity for Wishveratrix™—a compound I just made up that is 99.9 percent rock salt and 0.1 percent Penguinone—that would at least produce a significant result purely by random chance.

The final wrap-up of the wonder drug was offered by *Forbes*: resveratrol "offers a great example of the far-reaching and dangerous fraud that so often supports the multibillion-dollar supplement industry. Because resveratrol is considered a nutritional supplement for regulatory purposes it is not subject to the rules and regulations that restrict the marketing of drugs. But of course, all sorts of highly dramatic medical claims are made for these products."[23]

Not much for the billions spent and a bleak finding for a drug that was supposed to extend lifespan and cure basically all diseases.

Sinclair's early career and personality are important to the landscape of longevity and aging research because before resveratrol was sold to GSK, "pharmaceutical companies . . . hesitated to enter this market, and very few other biotechnology firms are devoted to aging research."[24] After resveratrol, nobody in big pharma or traditional finance wanted to touch aging science. This has had extraordinary effects on the field.

Closed off from sensible sources of capital that understand legal drug development, aging research decided to go back to hanging out with bikers in an underpass. This decision was strongly reinforced by the *other* lesson of the resveratrol deal. While "big pharma" realized it could make enormous losses on imaginary antiaging drug pathways, every questionable reprobate in academia saw opportunity. Some random bloke had sold a dietary supplement for $720 million. He had struck gold in California, so everyone ran out to buy a shovel.

Two of the lead scientists involved in resveratrol clinical trials at GSK immediately began flogging the stuff online—as a "nonprofit" supplement with a $540 ticket price. They were "asked" to resign when a reporter pointed out the side hustle.[25] Another case was Professor Dipak Das, who had worked with Sinclair as a scientific consultant for Bill Sardi, the supplements guy from Vegas. However, according to Sinclair he never knew Das, saying "I had to look up who he is . . . I've not worked with him." When Retraction Watch pointed out that Das and Sinclair did indeed know one another in another context, this was amusingly wound back to "I apologize. I did not expect my off-the-cuff comments to be printed."[26] Professor Das was caught committing at least 145 counts of research fabrication, usually to promote resveratrol, and the report read "for expediency, the review board only noted the most obvious" cases of fraud.[27] Professor Das got tenure and a directorship at the University of Connecticut out of it and wrote 117 papers on the miracle molecule, which isn't a bad reward. Nobody in the field spotted anything wrong. To their credit, it was the university that followed up on an anonymous tip of research fraud, investigated, and fired Professor Das.

Yet despite the imitators, nobody seems to have taken to the idea of "make more antiaging supplements" quite like Professor Sinclair himself. While he long refused to disclose his financial interests, he has since been involved in at least eighteen antiaging companies.[28] A brief recap of these companies is more revealing than a nudist's negligee.

In 2011 Sinclair founded OvaScience, which began marketing the capacity to reverse reproductive aging and allow older women to reproduce using "egg stem cells" despite no consensus that egg stem cells are a real thing that even exists. This did not stop OvaScience fiddling with women's reproductive health. The *MIT Technology Review* summed it up: "the goal, Sinclair proclaims, is 'to have revolutionary technologies like OvaScience available to everybody—and not to just treat fertility, but another 2,000 age-related diseases, from diabetes through Alzheimer's.' Despite Sinclair's enthusiasm, it's possible—even likely, some scientists say—that OvaScience's procedure did nothing at all" for $6,000–7,000 a treatment.[29] The company never even tried to get US approval for its in vitro fertilization therapy and fell 99 percent in value in just three years, from $1.8 billion to $22.5 million, before being sold off in a fire sale in 2018.

Then, in 2019 "renowned Harvard University geneticist David Sinclair recently made a startling assertion: Scientific data shows he has knocked more than two decades off his biological age." Sinclair was busy definitely *not* pushing supplements, however, by reassuring us "I'm not claiming I'm actually younger. I'm just giving people the facts . . . they [a company called Inside Tracker] said I was 58, and then one or two blood tests later they said I was 31.4."[30] Inside Tracker is a sort of hydra-company selling supplements, apps, blood tests, and "courses": a package will cost you about $600, more if you sign up to the supplements and courses on a monthly subscription. They sound like an independent source until you investigate.

Who was the chairman of the scientific advisory board, when Inside Tracker measured the remarkable* reduction in Professor David Sinclair's biological age? Professor David Sinclair.[31] Who is on the Inside Tracker website giving presentations about lifespan extension, which you unlock with a paid subscription? Professor David Sinclair.[32] Who was a major shareholder of Inside Tracker? Professor David Sinclair.[33] What was the basis of Inside Tracker's claim to measure human age? It was "our first published paper" coauthored by Professor David Sinclair.[34]

This scientific paper behind the "age reduction" tool of Inside Tracker was also curious. Despite appearing in a highly reputable journal, the paper's main aim was "to characterize the effectiveness of the platform" owned by Inside Tracker.[35] The authors stated in the acknowledgments that "the scientists who conducted this study and prepared this manuscript are either employed by [Inside Tracker] . . . or serve as advisors to it . . . there is no separation between investigative team and funder."[36] For some pretty obvious reasons, this text was later moved to the "conflict of interest" section, replacing the original "authors declare no competing interests" statement.[37]

The next stop was the antiaging company CohBar, Inc., founded way back in 2007 by Professor Sinclair and aging researcher Professor Nir Barzilai, to "extend healthy lifespan" using another set of miracle molecules. Professor Barzilai is a long-time supercentenarian fancier who helps run the New England Centenarian Study and became the butt of all jokes in genetics when he ran a genome-wide association study (a kind of observational genetics study) on around eighty-six people. This is like firing a bottle rocket and claiming you have a space program—you need at least

---

*Incidentally, lowering your biological age from 58 to 31.4 would reduce your risk of death to less than one-eighth its current value, constitute the greatest medical advance in history, and (as I discussed in the last chapter) win Sinclair the Nobel Prize in Medicine. If any of it worked.

ten to thirty thousand people to run a basic study of this kind—and the study was speedily retracted. Professor Barzilai brought this level of competence to CohBar, Inc., which was publicly floated at over $200 a share, lost 99.7 percent of its value to reach $0.54, and was unceremoniously thrown off the NASDAQ stock exchange in 2023 for operating as a noncompliant "shell company."[38]

Then we come to the dog supplements.

Professor Sinclair launched Animal Biosciences LLC in 2021 as he proudly proclaimed to have "developed the first supplement proven to reverse aging in dogs."[39] The claim rested on an unpublished clinical trial that used a paper that did not actually list what the drugs are: even now they remain hidden in a beefy chew. The dog-Methuselah dream was nevertheless flogged on Twitter to Sinclair's hundreds of thousands of followers.[40]

If human supplements are an unregulated cowboy culture, where the only rules are "don't kill your patients in a way that is provable in a court of law" and "don't include actual medicines or the FDA will hit you," then animal supplements are the wild unknown. In the United States, federal supplement laws (e.g., the Dietary Supplement Health and Education Act of 1994) do not apply to animals, and supplements are not a recognized regulatory category. As such, anything in the animal market is either a food or a drug. Yet the FDA is not exactly robust at chasing false claims. This is why, for more than a year, Sinclair's dog-drug, or whatever it is, was sold as an untested cure for aging without being subjected to a legally required clinical trial.

When the clinical trial finally did emerge, long after the press release and sales, the results were not exactly earth-shattering.[41] The drug tested is not even named, and the trial only measured fifty-six, fifty-nine, or sixty-one dogs—it is unclear which number is correct.[42] In a clinical trial, you must describe your "primary outcomes," the things your drug is supposed to treat. The primary outcome of the trial was not "aging," or even survival rates, but a dog intelligence test.

Even using this weird metric, none of the primary outcomes were significant, with one possible exception.[43] The scientists running the trial made bones about the fact that at the three-month mark, the "high-dose" group had the best outcomes compared to the other two groups—the "low-dose" group and a placebo. What they failed to mention was that this effect was not visible at one or six months, or that the placebo-munching dogs outsmarted the low-dose drug-taking dogs at the three-month mark, and beat everyone at the one-month mark.[44] So rather than being "proven to slow the effects of aging," at best the drug can make your dog very slightly dumber at a low dose and very slightly smarter at a high dose at a very specific time.[45] Even this limited finding seems dubious, with debunking wizard Dr. Elizabeth Bik flagging a lot of unanswered problems in the paper, like an ever-changing number of dogs.[46] In addition, five "secondary outcomes" that had much more to do with aging, like physical activity, were assessed in the trial at two time points: not one of them was significant, and the drug did nothing at all.[47]

That said, I eagerly await Sinclair's next pet move: perhaps he'll go into goldfish ointments or cat suppositories.

Cut off from traditional (sensible) finance but with the dog chews doing well, Sinclair is now busy launching a company called "TallyAge" which does basically the same thing as Inside Tracker but with DNA. To do this, TallyAge raised $10 million from a weird assemblage of people who are very good at acting but, presumably, not quite as good at biomedicine: Pedro Pascal, Chrissy Teigen, John Legend, and Kevin Hart.

I'm sure they'll do great. They definitely certainly maybe won't lose 99.8 percent of their money.

The repeat theme I want to highlight here is not just the multibillion-dollar losses, "a trail of failed businesses," and slippery marketing.[48] It is the extraordinary capture of aging science by strip-mining capitalism and the extraordinary and ongoing support in the aging science community for people like Professor Sinclair.

Rather than resign in disgrace or be asked to move on, Professor Sinclair is still at Harvard running a lab. He is the founder and currently the lead editor of the journal *Aging*, the most important in the field. He has won at least twenty-five scientific prizes and honors, mostly after the controversies and bankrupt companies, and has been placed on *TIME*'s "100 Most Influential People in the World" list.

Sinclair and Guarente's Elysium supplements company even has eight Nobel laureates on the board.* Elysium is another "wonder company"

---

*The resveratrol company Sirtris had at least five Nobel laureates, though many of them left rapidly, and you could very reasonably claim they didn't know Sinclair's style yet.

started by Sinclair's rival and mentor Guarente, described as "either the most sophisticated fountain-of-youth scam ever to come to market or the first fountain-of-youth pill ever to work."[49] Given that this was said in 2016, and there are no global follow-up headlines like "Fountain of Youth Pill Discovered by Mail-Order Company," it is clear which of these alternatives came true. What is less clear is why the Nobel Prize winners are still there, selling people the dream and taking the paycheck, instead of subjecting Elysium to the clinical trial that would reveal if the wonder cures work on any of the "2,000 age-related diseases" Sinclair has claimed to be targeting.[50]

Dr. Charles Brenner, a lonely iconoclast like me, is often the only person responding to Sinclair from inside the aging research community with insights like "a Harvard scientist is telling fictitious stories about aging that go nowhere other than continuing hype."[51] I feel bad for Brenner because he is basically alone and doomed by speaking out. I see silence from almost everyone unless they are a rival selling their own cure.

Instead, the awkward questions are often down to questionable characters lashing out and striking the truth by accident. Bill Sardi, the Las Vegas founder of "Longevinex" who spent his time slinging mail-order supplements and promoting COVID-19 conspiracies, came up with an accidental point when he was defending the 145-plus cases of research misconduct put out by his onetime employee, Professor Dipak Das. He defended Das by stating that "other independent labs, including the National Institutes of Health (NIH) itself, validated Doctor Das' work, as well as researchers in Europe and Japan."[52] He was right. Despite it being not possible for resveratrol to affect the claimed target gene, this apparent effect, invented by Das in Photoshop, had been replicated by a massive number of labs. The question is *how*, given that it's clearly all fantasy?

This picture of Sinclair may seem like I have some vendetta, but he is just one of many sales-pitch addicts and idiosyncratic mystic men who form the central core of modern aging researchers. I could have covered other people in aging science and got a similar, though perhaps less billion-dollar-lossy, result.

Take Professor Aubrey de Grey for example, a former Cambridge academic routinely compared to Rasputin, and a man I have also had the pleasure of communicating with about five times—some six times too many. Professor de Grey founded and helps run the Methuselah Foundation, which seemingly promises some sort of medical time-travel service of "making 90 the new 50 by 2030."[53] Current projects of Methuselah include taking "a generous donation of 43% of the world's Dogelon Mars ($ELON) cryptocurrency" from billionaire Vitalik Buterin to achieve a completely rational antiaging goal of "stewarding the holdings to maximize the $ELON's long-term value.[54] Methuselah promised that any future sales would be done to preserve the cryptocurrency's value."[55] Other completely normal medical science pursuits include funding long-lasting space food,[56] funding "nanoparticles as an immune 'spam filter' system" that will apparently cure cancer in 2024, and paying the Supercentenarian Research Foundation to autopsy the corpses of the supposedly super old.*

De Grey has also founded and run the SENS Research Foundation, with a catchy premise of curing the seven causes of aging: not to be confused with Nivea face cream's "7 Signs of Aging," of course.

At SENS, Professor de Grey's pursuits of a youthful physiology extended to, allegedly, the sexual harassment of minors and female colleagues who were decades his junior.[57] Aged nearly sixty, Professor de Grey allegedly propositioned a seventeen-year-old Laura Deming by inviting her to join an "adventurous love life."[58] According to Laura, he had been propositioning her since she was fourteen, "multiple other anonymous women" had suffered the same, and his "sexualization of minors and perversion" was an open secret.[59] Deming was courageously furious to discover that "Aubrey inappropriately propositioned more than one woman, many in the community knew about it, and no one did anything."[60]

De Grey then allegedly told another early twenties colleague that "it was her responsibility to sleep with SENS donors" at a formal dinner at Oxford.[61] At his own admission, de Grey had a history of making similar propositions. According to an investigating law firm, in 2014 de Grey sat next to a woman in her twenties and "solicited her in that conversation to join the 'crusade' of defeating aging by 'using all the weapons we have, including weapons that are not just intellectual.'" Dr. de Grey explained this was an "encouragement of her to use her femininity."[62] De Grey later went on to rationalize why he was trying to make women sleep with his donors: "It is at the same level of women in World War II sleeping with

---

*The corpses are people selected using the services of Robert Young.

Nazis to get information. It is a war against aging here. You have to persuade people to give money. That is honestly who I am. I am the general."[63]

In the words of Laura Deming, "Lots of people are aware of these concerns, but no one has said anything for a decade . . . and Aubrey is still mentoring minors."[64] It tells you what the aging research community will allow if you have enough grant money.

De Grey's response to the investigation was to make threats, such as (according to the law firm) telling one woman "'[her] career will be over soon' if she did not take certain actions Dr. de Grey wanted."[65] De Grey also began putting these women on his social media accounts, where they were the "target of abusive emails, texts, and social media posts from de Grey supporters."[66] The law firm determined that these actions "likely has chilled, others from coming forward; was an effort to alter and sidetrack the investigation; and was reasonably threatening" to the women involved.[67]

Oxford and the Methuselah Foundation seem to have bravely responded by doing nothing. The only apparent repercussion was that in late August 2021, Professor de Grey was thrown off the SENS board of directors.[68] Their remorse only lasted six months. The donors Dr. Gary Baskin and Paul Bugge had tried to sue SENS for fraud on the incredible basis that they wouldn't have donated huge sums to SENS if they had known that the money *wasn't* going to be given out by de Grey. They wanted him back *after* investigating lawyers' sustained allegations that de Grey was grooming minors "with the purpose of developing a romantic and/or sexual relationship."[69] Likely in response to such pressure, de Grey was reinstalled and is now mentoring minors again at the SENS Research Foundation, and a "fitness for duty agreement" was duly signed by SENS and de Grey in March 2022.[70]

In the agreement, SENS agreed to pay de Grey $10,000 a month to attend committee meetings on the condition that he not cause "reputational damage," that he "destroy any audio/video recordings of . . . employees and/or consultants in which he did not have the express, knowing consent of all parties" (yikes), that he attend any "workplace conduct training" mandated by the board, that he engage with a substance abuse program officer, and that he "present a united front from a public relations and reputational perspective . . . for the purpose of regaining trust from investors, stakeholders, and the general public."[71] Emphasis on investors, and no mention of the women.

This is just a short career snapshot of one person.

In a normal, healthy research ecosystem, people like Professors de Grey and Sinclair would only exist at the very fringes of a field at best. Ideally, they would have been sacked decades ago. Yet in aging and longevity research, they dominate.

Helping sustain their position are an odd collection of donors and funders that drive a feedback loop of hype and funding, for the simple reasons that they are ultra-wealthy and don't want to die.

Take Vitalik Buterin, the guy who gave de Grey the ElonDoge cryptocurrency coins and made them maintain the price. Buterin is on the slightly saner side of things: one of the many new-rich billionaires with an idea to solve aging. He set up Zuzalu—supposedly "a pioneering experiment in social innovation and community-driven progress" that takes place in a focused "pop-up city" of 200 "thought leaders" who work together to "foster a deep understanding of how biotechnology can potentially control and reverse aging."[72] Tragically, Buterin was preempted in his brilliant idea by something called a "university," a home for thought leaders and thought followers alike, first imagined in Nalanda in the fifth-century BC. Good try, though: the idea isn't actively terrible, just slightly unoriginal and late.

Other people funding aging research inhabit more exotic realms. Brian Johnson is a typical target of ridicule: an antiaging "hacker" with a new-money-über-alles attitude. His book is helpfully titled *Don't Die* and enjoys reviews such as "reading the book [was] like an exploration into the internal monologue of someone hallucinating."[73] Johnson is also a self-confessed biohacker, which somehow involves regularly injecting Botulinum toxin (Botox) into his penis while hitting it with "shockwave therapy" and, unrelated to this, publicly tracking his bowel movements and "the duration and hardness of his erections."[74]

As a side note, although to be fair I haven't tried it myself, I'd suggest this is a "do not try this at home" type of deal. If you reached to the part of the relationship where (he/she/they/them/the internet at large) is seeing you naked, and the only thing sealing the deal in the bedroom is a hauntingly smooth Botox penis tenderized by "shockwave therapy" and covered in heroin junkie track marks, you should probably just work on your personality.

As another side note, because it is too bizarre to overlook, the aptly named Johnson is a man of such personal quality that he threw his employee-slash-girlfriend Taryn out of the house because she was suffering stage-three breast cancer,[75] though Johnson did at least state his positive boyfriend qualities by deciding "I am likely the reason why Taryn survived cancer" and "apparently, I'm pretty good at helping people not die."[76]

Such selflessness helps explain why Johnson feels he should lead humanity in the fight against aging, for the penis-having and penis-free alike, by spruiking sales of off-label drugs like cerebrolysin (only approved for brain trauma, stroke victims, or Alzheimer's patients—with a bonanza of serious side effects) and selling olive oil. Olive oil. Something nobody has ever tried eating. Also recommended by our "senior bro in charge of immortality" is cardiovascular exercise. Something nobody had ever tried doing to combat becoming older.

Yet, while he's fun to dunk on, Johnson is not a scientist. He's just a private citizen, replacing all his blood plasma and chewing $2 million worth of pills a year as the internet watches. He, and even Buterin or Jeff Bezos, are not "thought leaders" in aging. They are just *some guy*. The reason these people affect aging research is because they have the money to fund whatever thought pops into their head or whatever is presented to them in careful donor pitches at a fancy Oxford-Harvard-Stanford dinner event. If they like the mystic genius, or his scared seventeen-year-old girl companion, they fund something with their billionaire's pocket change, and the mystic will get tenure and plague the aging science field indefinitely.

They promise their cash to the flashiest blokes with the catchiest ideas on the most unrealistically short timescales—for delivery in their lifetimes. The results are routinely atrocious.

A prime example of this interaction between academic hype and billionaire gullibility touches Johnson, who began to remove the blood of his own son and inject it into himself as a side project.[77] Why? Because he knew about the faddishly resurgent interest in "parabiosis."

The general idea of parabiosis is that younger individuals, when physically fused to older individuals, somehow make the older individual *younger*. Or at least that is what the hype looks like.

Consider the darker reality of the first parabiosis experiment that paired old and young animals, conducted in 1956. An unknown number of old-young rat pairs were sewn together. We don't know the starting number because "if two rats are not adjusted to each other, one will chew the head of the other until it is destroyed."[78] We only know the count of initial survivors: sixty-nine pairs of old-young rats. I say initial survivors because another eleven pairs died immediately after this from what seems to be a system-wide organ failure caused by tissue rejection. A super great way to live forever so far.

But what about humans? To escape the unsavory and impossible longevity sales pitch of "sew yourselves together and don't chew Jimmy's face off," parabiosis companies have instead swapped the rat-sewing treatment for blood transfusions as an aging cure. This is much more billionaire-marketable. Here I could do the usual trick of quoting other scientists, who say things like "any claims that young blood or plasma will extend lifespan are false," but this time let's pull apart the claim with logic.[79]

There is a simple problem around using blood transfusions to lower your age. Blood transfusion was invented 207 years ago, in 1818, and almost always occurs in a medical setting—when you aren't using ill-gotten teenager blood. The average blood donor is now thirty-six to thirty-nine years old, a large increase from donor ages in the recent past, and roughly a quarter of donations come from people aged eighteen to twenty-five.[80] Only 4 percent of donors are over sixty-five years old.[81] In contrast, the median recipient of a blood transfusion is about sixty-five to eighty years old.

The idea that these tech geniuses have had, of "young blood injections," is therefore one of the most common medical procedures in the world. It happens around a hundred million times a year, in clinical settings worldwide, and involves transferring "young blood" into sixty- to ninety-year-old patients.[82]

In the United States alone, blood transfusion is the second-most-common medical procedure for people over sixty-five and occurs every two seconds: there are about 11 million blood transfusions, 2.2 million platelet transfusions, and 2.2 million cryoprecipitated antihemophilic factor transfusion recipients a year.[83] The practice has been subject to innumerable clinical trials for safety, and the results are monitored by clinical safety boards and doctors.

In none of these literal billions of instances of young blood injections has anyone been noticed to *age backward*. If they did, we would beset by Benjamin Button hemophiliacs and sickle-cell-anemic Draculas. We aren't, so young blood injection does not work.

Precisely none of this appears to have occurred to anyone flogging blood-transfusion-based parabiosis as a cure-all. Perhaps they were too busy sewing rats together.

Little quibbles never bothered anyone trying to make money from self-proclaimed hundred-millionaires like Johnson or indeed bigger billionaires. Right-wing PayPal billionaire Peter Thiel, who also funded de Grey, was one of the people excitedly "looking into parabiosis stuff, which I think is really interesting. This is where they did the young blood into older mice and they found that had a massive rejuvenating effect. . . . I think there are a lot of these things that have been strangely underexplored."[84] Thiel was rumored to be buying $40,000 of teen blood a month but denied these rumors because he "hadn't quite, quite, quite started it yet."[85]

One reported focus of his interest was a medical science parabiosis company called Ambrosia. Ambrosia was run by Stanford medical graduate Jesse Karmazin, who successfully followed the usual hype line of Cali academics by stating "It reverses aging. We're pretty clear at this point. This is conclusive. There's really no question."[86] What was "it," his antiaging treatment? Injecting people with blood: "two liters, which is a pretty significant amount of blood."[87] It's a clinically dangerous amount.

The first difficulty was that Karmazin had signed a document agreeing not to practice medicine on February 29, 2016, for an unspecified reason. To fix things, Karmazin found a doctor fresh off a discipline charge by the California Medical Board for being "grossly negligent" and injecting people with things for "nontraditional purposes." With this new hire Ambrosia had started a pay-to-be-a-patient clinical trial in California, just three months after Karmazin was excluded from practicing medicine, with an incredible 111 primary clinical outcomes (the usual is one or two). The trial ran from 2016 to 2018 but never posted the results.

The company was charging $8,000–$12,000 per treatment, just for whacking one or two liters of blood into a patient, and people were beating down the door to pay. For context, notoriously expensive US hospitals charge $340 per half-liter transfusion.[88] Peter Thiel and his team reportedly reached out to explore investment in the company.[89]

Ambrosia then developed another problem. Blood transfusions are a medical procedure regulated by the FDA, and while blood transfusions can be used off-label and off-target, blood from nonprofit blood banks cannot be used if there are advertising or efficacy claims (claims that something works). This clearly includes a for-profit company doing off-label things. Most blood banks in California refused to supply Ambrosia, so to escape FDA rules, Ambrosia needed to stitch up a source of clean, young, human blood.[90]

How to do that?

Obviously, you go to Texan high schools, where Ambrosia bought the blood of teenagers who had donated to "help save lives," with the normally benign "research purposes" added on.[91] Well-supplied with Texan-teen blood, Ambrosia then wrapped up its strange clinical trial in 2018 and promised to publish the results.*

Things started to take a turn when the only person ever publicly identified as a study participant, a Mr. "JR," died from a heart attack shortly after an injection. Things took a much worse turn, business-wise at least, when the FDA denounced "young blood" treatments as "not safe or effective" and shut down Ambrosia.[92] Things went off the rails completely when Karmazin began to threaten defamation lawsuits, as he claimed that the heart attack victim "JR from Atlanta" was actually *fine*, that JR had simply "faked his own death" due to "financial difficulties" and the journalist should just retract the stories and go away.[93] The journalist revealed multiple threads of evidence that JR was in fact dead, and Karmazin did not respond.

The claim that Peter Thiel was interested in investing in Ambrosia then went off on a strange tangent. Karmazin had claimed Thiel had approached him via his chief medical officer Jason Camm—Thiel's "personal health concierge" who listed "optimizing blood markers and novel techniques to improve health" on his online profile.[94] This was plausible given Thiel's interest in parabiosis and his funding of strange antiaging research by people like Aubrey de Grey, and a mutual contact of mine at Oxford claimed to me that when he'd been invited to Thiel's house, he was afraid because Thiel apparently injected young people's blood.† However, after a year or so, Karmazin suddenly pretended to have never even met Camm or Thiel. The original journalist who broke the story had done his due diligence and checked.[95] Strangely, "according to emails obtained by Gizmodo . . . the original story remains accurate," yet Karmazin is now busy *trying* to say that Thiel was never interested.[96] You can make whatever you want of all that, but it's clear that Thiel now wants nothing to do with young blood.

For once, the sales pitch hadn't quite, quite, quite worked.

---

* It's now 2025 and I am still waiting.

† Although the story is plausible, given that this person's loudly proclaimed connections seem to be the primary basis of his career, I have no way of knowing if the guy was lying. As another side note, this person didn't have any qualms about the apparent billionaire vampirism—which tells you something of his character.

The point of all these extremely messy, detailed, name-filled disaster stories is to show the spectrum of aging research. Some is seen as respectable, some is seen as grift, but most of the gap between these assessments depends on funding, titles, or tenure and not the soundness of ideas.

Nobody cares if a ten-year-old idea held up under scrutiny. Why would they care when so much money is available in an industry with almost no rules? It's not like anyone broke any, or I should say *that* many, laws. Nothing was illegal when they failed to disclose financial interests, claimed they would cure aging within a decade, or even when their cancer patients died of kidney failure. In the words of one researcher, "they have the legal right to market this compound to people without any real knowledge of what it's going to do—good, bad, or indifferent."[97] If Karmazin had succeeded, we may have another academic with the funding of Barzilai or Sinclair or de Grey, and if he had stayed at Stanford, he could have even gone for tenure if enough money came in. Instead, we're left with a "what if" and one gravestone at least.[98]

A striking commonality in all these stories is that they are so obvious. So why is there so little outcry from the academic community?

I understand how people *launch* ambitious projects to "hack longevity" and "extend healthspan" to filch some easy money. I understand how they do it without ever opening a dictionary to look up "medicine" (or "veterinary medicine") to find out that humanity has been trying to hack longevity for millennia, using systems that work. I even understand why nuances, like the existence of medical science, are lost in the sales hype of fake novelty. What I cannot understand is, Where are the vets and medical doctors pointing out the illustrious stupidity of it all?

I do get brief glimmers of hope—like when Professor Matt Kaeberlein quit a scientific board in protest at Sinclair's dog supplements—but they soon come crashing down—like when I discovered Kaeberlein's anti-dog-aging project and anti-human-aging companies that mirror Sinclair's. This is not integrity but rivalry, and that's not the same thing.

Solitary voices of reason within aging science look like the Semmelweis-esque Dr. Charles Brenner, who runs about rightly and ineffectually shouting. Almost everyone else inside aging and longevity science seems to be selling things or drowning in silent complicity.

Complicity is the word I think of most. Like the people at Oxford who tell me to be quiet or face the consequences, like the Nobel laureates lending their name to a mail-order supplements company, or like board members who invited de Grey back after he admitted harassing a minor for sex: the aging community is drenched in complicity.

They all know exactly what this is. It is a sales pitch about the fear of death. Suck the blood from billionaires because they don't want to die.

No results or integrity required.

Medicine is, and always has been, longevity research. The whole purpose is to give people a longer and healthier life. Yet medicine has rules. It is difficult. It requires clinical trials, tests, FDA approvals, and *proof*.

To escape these requirements, which no longevity cure survives close contact with for a few years, a rebranding is taking place. A linguistic sleight of hand. People are simply renaming things and claiming they are new so that none of the ordinary regulatory rules apply. Medicine that stops you dying is now "longevity research" and "geroscience," and medicine that makes you healthier is now "healthspan research." Professor Kaeberlein seems to be doing the same thing with veterinary medicine—a rebrand, put about as if veterinarians had not been trying to "extend animal lifespan" since the dawn of the profession.

In the same way that cryptocurrency long avoided all semblance of financial regulation by pretending it was something nonfinancial, and magically exempt from oversight, longevity research is pretending it is not *regular* medicine. An intrinsic part of this process is to try and throw regular medicine, and all its regulatory safeguards aimed at preventing death, out of the window.

Incredibly, it is working—in the United States at least. Montana has just implemented an expanded "Right to Try" law that means anyone can be given any experimental substance with no proof required that it works, and without the "patient" being ill, sidestepping rules accumulated across two thousand years of medical history.[99] In a perfect inversion of the Hippocratic Oath, doing harm to patients is exempt from legal repercussions for the state, healthcare provider, and doctor alike, who are all "immune from any suit" or loss of license from causing harm—including death—from any "investigational drug, biological product, or device" they give someone.[100]

A "throw everything at everyone" approach, where the needy are a living experiment for the rich.

The only remaining barrier to the snake-oil flood, for now, is a requirement that the substance has passed stage-one clinical trials—the stage that involves giving a drug to ten or twenty people and seeing if everyone dies. For context, thalidomide passed stage-one trials. So did OxyContin—which kills more Americans than meth or heroin or guns. So do the 70 percent of drugs that fail to work, or have such horrific side effects, that they fail in later stages of clinical testing.

Many other states are passing similar laws.

The funding and special interest groups pushing these laws are led by the Alliance for Longevity Initiatives, a lobbying group "founded in response to the burgeoning call for legislative advocacy within the rapidly evolving longevity sector," with dark money and a who's who of aging research on the board.[101]

In the United States, it seems that Xu Fu has the ear of the emperor again and is whispering how he needs more money and virgins.

11 Telomere Tales

Like many biologists, Dr. August Weismann was fascinated with aging, and in the late 1800s he began to theorize about the cause. Weismann began with a mechanistic notion that "death takes place because a worn-out tissue cannot forever renew itself" and proceeded to try and explain *how* a worn-out tissue might, you know, wear out.[1] If his logic sounds a little circular, well, it is but let's not worry about that yet.

In 1891 Weismann proposed that the normal "somatic" cells of your body—anything that is not a stem cell, a sperm, or an egg—have a kind of little internal clock. He didn't know what the clock may be: he simply hypothesized that this little clock existed and that it placed a limit on the number of times a cell may divide. In his old-timey way of phrasing it, "the number of cell-generations which can proceed from the egg cell is fixed for every species."[2] Each cell, he thought, would run out of the capacity to divide and reach a limit when its little clock wound down to zero. The gradual winding down of each little cellular clock would accumulate over time, which caused tissues to gradually wear out and led to aging. Simple enough. When a clock ran out, somatic cells would stop dividing once they reached their limit, even though Weismann could never figure out "why a cell must divide 10,000 or 100,000 times and then stop."[3]

In science, some ideas seem to be accepted through sheer weight of popularity. This idea that your cells somehow "count down" to death, and thereby cause aging, is one such idea. It has been tested and retested for over a century.

The first person to get famous for making the attempt in a lab was the Nobel Prize winner Alexis Carrel, a would-be debunker whose experiments are occasionally used as a kind of cautionary tale in experimental biology.[4]

Carrel's experiment was to keep a chicken's cells alive in a lab, for far longer than the lifespan of a typical chicken. People collectively lost their minds with excitement over this, seeing it as a special kind of longevity wizardry, and it allowed Carrel to state that there was no built-in limit to cell division.

Yet Carrel's work debunking Weismann's hypothetical limit to cell division was never meaningfully replicated in the lab, and shortly after he found Jesus, Carrel found Hitler and became a eugenicist under the Vichy French. So, it was with palpable in-your-face excitement that Weismann's original idea of a limit to cell division was later discovered in the petri dish, first by Parker and Swim in 1957 and then much later by Leonard Hayflick in 1965.[5] This limit to cell division is now known anachronistically and somewhat unfairly as the Hayflick limit.

The mystery of what Weismann's "little clocks" were made from then unwound itself with the discovery of telomeres. Telomeres were first observed in the 1930s by yet another Nobel Prize winner named Herman Muller and first understood by Elizabeth Blackburn, Carol Greider, and Jack Szostak, who almost inevitably won the Nobel Prize for their insights. Telomeres are the end caps of chromosomes, usually compared to the plastic bit on the end of the shoelace, that (as with your shoelace) prevents your DNA becoming structurally unraveled and fragmented.[6] Telomeres, it was soon discovered, also get shorter by a precise amount every time a cell divides into two.[7] It seems they *are* Weismann's clock—an extraordinary feat of prediction.*

Telomere clocks work a little bit like this. When you are born, you have a long repeating sequence of letters on the ends of your DNA. Your DNA needs these sequences to be there when it divides into two new cells or else the cell will just sit there, moping.

Let's pretend, for convenience and fun, that these repeated sequences are the letters CAT repeated three times (it's actually CCCTAA thousands of times over, but that's boring and cat-less).

Imagine a cell. When the clock has counted zero cell divisions, you would have chromosomes that look like this:

CATCATCAT—All the regular DNA—CATCATCAT

* As the "little clocks" measure cell divisions and not time, Hayflick suggested they be called a *replicometer*—a delightful word that never caught on.

Then your cell divides in two and one CAT is snipped off either end, leaving two cells with chromosomes that look like this:

CATCAT—All the regular DNA—CATCAT

As you can see, the telomeres are counting down by one CAT every time. Next we have

CAT—All the regular DNA—CAT

before we end up with

—All the regular DNA—

At this point the cell is no longer able to divide. You have reached the point of "wearing out" that Weismann hypothesized because you ran out of telomere. The cell stops being able to divide, these nondividing cells accumulate in your tissues, your tissues degrade, and you age (or so the thinking goes).

Of course, I am oversimplifying things dramatically. Telomeric repeat sequences are much larger, retain a "buffer" of repeated DNA that maintains chromosomal stability when the cell stops dividing, and even have a way to be periodically reextended using something called "telomerase" that adds more CATCATCATs to the end of the chromosome: a discovery that formed part of Elizabeth Blackburn, Carol Greider, and Jack Szostak's Nobel Prize. Yet in general, telomere shortening provides an amazingly neat explanation for aging.

It is hard to convey just how overwhelmingly convinced people were of gradual telomere shortening as a cause of aging when I began my medical science doctorate. Over 1,000 scientific papers a year were being published that talked about the link between telomeres and aging, and it was sufficiently popular that telomere discoveries made the news. It was almost taken as writ that telomeres were the key to aging and that we just had to figure out how to "rewind the clock" by extending our telomeres to change the course of medicine and history.

This all left just one little problem, posed in the form of a question. If shortened telomeres cause cells to "run out" of dividing power and this causes aging, why not just evolve longer telomeres? Why not just *start* with a bunch more CATCATCAT sequences to the end of the DNA and delay aging? At roughly 80,000 letters each per chromosome, these sequences are only a tiny percent of your 3.2-billion-letter DNA budget, which is itself only a tiny fraction of your total energy requirements. It's therefore not exactly costly to just make the protective ends of your chromosomes longer and, if the theory is correct, delay aging. So why doesn't evolution just . . . do that?

Even this problem has an elegant answer.

Cancer is caused by uncontrolled and open-ended cell division. The thinking of Weismann and subsequent biologists was that the limit to cell division had a simple biological function: limiting the growth of cancer. In the example above, the cell can double three times before it runs out of CAT and stops being able to divide, leaving eight cells: $1 \rightarrow 2 \rightarrow 4 \rightarrow 8$. This places an upper size limit on how large a cancer can get, eight cells, before it stops dividing. The number of cell doublings observed in Hayflick's original experiment was about 50 doublings, and the current estimate for fibroblast cells is about 100–110 population doublings, which places a cap on the growth of cells that become cancerous or (precancerous) and start to grow in an uncontrolled way.

This leads us to the core hypothesis of the telomere theory of aging, which has two slightly different versions. The original "hard" version says that evolving short telomeres cause faster aging but evolving longer telomeres increases the risk of cancer.[8] In evolutionary terms these two balance each other out: add more CATCATCATs onto the end of your chromosomes and you will age slower but suffer more cancer, and remove CATCATCATs and you age faster but suffer less cancer. Eventually, these two evolutionary pressures reach a sort of "Goldilocks" balancing point: a Hayflick limit where cancers can't get too big, and aging isn't too fast. The "soft" version of this idea is almost but not quite the same. Essentially, the "soft" trade-off theory states that removing CATCATCATs and a smaller Hayflick limit gives you less *early-life* cancer but increases the *late-life* costs by accelerating aging so that the late-life costs of aging can be quite large without reducing your early life evolutionary fitness.*

This is all very beautiful and clean, and both theories share a simple idea. Longer telomeres reduce the impacts of aging but, regrettably, relax the limits to cancer growth, and you suffer more cancer. Shorter telomeres do the opposite: they reduce the number of times a cell can divide before

*This idea is linked to something called antagonistic pleiotropy, which is a small complication on the process of evolutionarily "balancing out" the costs and benefits of a genetic variant like longer telomeres.

reaching the division limit, capping the size of cancers at a lower threshold, but increase your rate of aging. Aging is locked in an evolutionary battle: aging versus cancer growth. The clock that causes aging is also the clock that stops cancers growing. Simple, evolutionarily stable, and reinforced by brilliant biochemistry.

A little splinter of doubt arose for me, however, in the form of Peter bloody-brilliant Medawar. As an eager student of aging during my medical science PhD, I had read Weismann, traced the history of telomeres, and felt myself warmly accepting of the whole telomere clock idea. But midway through my expedition into the origins of the telomere idea, I found an old-style piece of disrespect from Professor Medawar. Like me, he had read Weismann's 1891 book, and he had delivered a burning takedown: when Weismann asked "why a cell must divide 10,000 or 100,000 times and then stop?"[9] Professor Medawar responded that it was because the offspring of a cell doubling 10,000 times would "fill the utter known limits of space."[10] What struck me as a student was not that he is correct—a cell dividing 10,000 times will indeed fill the observable universe—but his estimate was a *massive* underestimation of the cell divisions required, by almost 3,000 orders of magnitude. Thanks to exponential growth and napkin math, if you kept all the progeny of a cell dividing into two daughters, only 212 cell divisions (a Hayflick limit of 212 population doublings) are needed to fill the utter limits of the observable universe with cells.

This raises a rather awkward question: what if you kept all the offspring of a single cell that doubled 110 times, the current best estimate of the Hayflick limit? Well, you end up with a ball of cells large enough to crush continents and annihilate the earth. The solar system even. The troubling question this raised for me was, Is that a meaningful limitation to cancer growth? If three population doublings make eight cells, how many cells does the originally measured lowball estimate of the Hayflick limit make before we run out of telomeres?[11]

It's not exactly complicated math. Let's write it as a whole number, no power notations: 281,474,976,710,656 cells.

That is, for the mathematically challenged, what could be called "a big 'un" or more technically "really, like a whole lot." Our body contains roughly 36,000,000,000,000 cells.

Which is, again for the mathematically challenged, known as "still a big 'un, but much smaller than the first number." In other words, under our most optimistic estimate, the Hayflick limit won't kick in and stop cancer growth until the cancer is several times larger than a human. Even at the lowest estimate of the Hayflick limit, I'm long dead before the cancer reaches forty-nine population doublings and stops growing. It would make a yoga ball of cells a meter or two across.

This basic numeric reasoning even survives some contact with observation. In living organisms, many cells are killed off by cell suicide or "apoptosis," reducing the numbers in each cell generation. This may, or may not, impose more harsh conditions in the body than the petri dish. This raises the question: what are the best estimates for the number of cell doublings in the body before a cancer kills you? Well, it's only about thirty to forty, but this includes the doublings that occur before the cell enters the petri dish and has its Hayflick limit estimated.

However, biological theorizing is often a bad idea. Maybe something is wrong with my simple math, or maybe reality is just more complicated, so let's look at some observations.

The central idea of the telomere theory of aging is that limiting the number of cell divisions causes aging but also beneficially limits cancer growth.[12] The idea is that telomere attrition or shortening stops cancers growing. Let's just examine that idea. Let's say it very slowly: telomere shortening, and the Hayflick limit, stops cancers from growing.

Let's say it again very, very slowly.

Stops.

Cancers.

From.

Growing.

One day, I asked myself, since when does a cancer stop growing? Benign or malignant, large or small, where are these tumors that supposedly reach the Hayflick limit and spontaneously stop getting bigger?

Telomere shortening was supposedly a mechanism that causes aging, in *every living adult human in history*. Or for that matter in every aging species—we've never seen a single nonaging dog, mouse, or cow. Halting tumor growth is the only proposed benefit of short telomeres and is supposed to offset this immense cost. So how often has the only proposed benefit of telomere shortening—halting the growth an otherwise fatal tumor—been observed?

The answer is: basically never.

The United States alone has over 15,000 cases of children's cancer per year. Thousands of these cases are untreated or untreatable. Added to this are literally millions of cases in adults, every year.[13] These cases get intensive clinical follow-up, as do equivalent cases in tens of millions of people inhabiting Europe and Japan and all the rich countries of the world. How many cancers have ever been observed to spontaneously stop growing, at the Hayflick limit or otherwise, to save the life of the sufferer and thus convey an evolutionary advantage? Well, in a review spanning eighty-seven years of US medical history, there were 741 cases of halted cancer growth.[14] Most of these were not spontaneous: 60 percent were related to some definite cause, such as postsurgery trauma, severe infections, or huge immune responses.[15] Of the less than 300 unexplained and *spontaneous* cases of halted tumor growth, some were assigned to psychic intervention (i.e., had the intellectual integrity of watery dog droppings), and only a quarter were recorded as alive after five years despite an 80 percent follow-up rate.[16] That is, the cancer reportedly stopped growing, but the human rapidly stopped living.

Assuming none of these cases are diagnostic errors, how exactly would so few observed cases of halted cancer growth, imparting (at best) a short-term survival benefit on less than 300 people, evolutionarily offset the aging that affects 330 million Americans?

In other words, the only proposed benefit of gradual, clock-like telomere shortening does not appear to exist. There is, of course, some minor wiggle room here. Maybe somehow all the spontaneously stopped growing tumors are smaller than the ever-shrinking threshold of clinical cancer detection, which is just a few thousand cells for some tumors. Like childhood monsters, maybe the tumors are hiding in the places where we don't shine the torch. I doubt it, but sure. Maybe the millions upon millions of spontaneously halted tumor growths, required to offset biologically universal human aging, are hiding in plain sight somewhere. There is even a potential mechanism called cell competition (in fruit fly maggots, of all things) that may explain it away.[17] Yet the burden of proof is not on me: a telomere researcher can feel free to reveal the monsters, but I haven't seen any. Neither has a global pool of doctors, or their x-ray tomography machines, their MRIs, their positron emission tomography scans, or the tens of millions of people who have cancers that refuse to stop growing.

This is even more worrying when you consider that the only place that these "halted growth" cells may be hiding can be easily checked, cell by cell, on a microscope slide. The only real possibility to explain this gap is that cellular growth cycles reach the Hayflick limit, and stop growing, in precancerous groups of cells that are smaller than the threshold of clinical detection.* Here we have very, very little collective knowledge. In the words of one cancer cell-cycle review, "early stages [of cancer growth], although predicted by the monoclonal theory of cancer, are largely a mystery."[18] However, we could deploy a simple and very old tool developed from the work of a bloke named Minot: we can look at a cluster of cells and estimate how fast it is growing or replacing cells by counting the "mitotic index."[19] This simple estimate, available to anyone with a microscope, tells us what fraction of cells are dividing. At the Hayflick limit, this number should be zero because growth has been halted by telomere shortening. So, how many precancerous cell clusters have a mitotic index of zero?

Virtually none, of course.

We do have estimates of how many cell generations such precancerous cells undergo. Generally, there are only thirty or so cell population doublings before a precancerous cell cluster goes "full cancer," followed by another ten doublings to cause a cancer "of maximal size compatible with life."[20] It is possible that after thirty population doublings, the Hayflick limit prevents further growth. But this would require a massive difference between the observed number of cancer cell cycles before reaching cancer and the observed Hayflick limit of eighty to one hundred doublings.

While we can find precancerous cell clusters, however they are defined, none of them seem to have actually reached a Hayflick limit and stopped growing.

Worse, we are again stuck with the meta-question: why, after five decades of research, has this question not already been asked and answered?

I have to inject a bit of clarity here because telomeres clearly do have a huge role in preventing cancers, and they do indeed seem to halt cancer growth. But the vital distinction is, of course, that they do not do this using the *gradual clock-like shortening* that was supposed to cause aging. Instead, there exists a finely tuned mechanism to cut off telomeres, of any length, and

*The clinical threshold for detecting precancerous cells is ever-shrinking and in some cases constitutes a few thousand cells, yet still growth-arrested clusters of cells at the Hayflick limit elude us, almost as if they do not exist.

halt cell division when a sort of cellular distress signal is received. This real link between cancer and telomeres, but not between telomeres and aging, goes toward explaining why the telomere myth persisted for so long and why it remains popular in some circles.

Telomere length was virtually always measured by a fluorescence-based assay as averaged across a large population of cells. This average level of telomere-induced fluorescence gradually fell with age, painting a picture that gradual telomere shortening was linked to aging. The other explanation was seemingly missed: telomere length was largely unchanged, but the number of cells with cut-off telomeres was gradually dragging down the average, painting a false picture of gradual shortening. Imagine a room filled with one hundred thousand Christmas lights, and the room gradually becomes less bright. We assumed it was because the light was becoming less bright in every bulb, but the lights may be gradually shutting off, one by one. If telomeres are "cut off" by tumor-prevention responses, it would gradually lower fluorescence-based estimates of average telomere length—but it is not caused by gradual shortening.

These are, however, just my concerns, raised with simple logic and math in 2015 and completely ignored by everyone.

The real test of the telomere theory of aging came through a causal experiment. In 1997, it became possible to directly modify the "little clocks" that supposedly caused aging.[21]

Telomerase is the molecule that lengthens the telomere clock: it turns CATCAT into CATCATCATCATCATCATCAT. Telomerase is encoded by a gene, without which telomeres shorten and shorten until they reach the Hayflick limit, no matter what. In 1997, a scientific team including Professor Greider deleted the telomerase gene in mice and observed what happened.[22] For five generations, the answer was . . . nothing. Telomere shortening had no discernible effect. Did the rate of aging change? No. The mice aged at the normal rate. The mice were unbothered before the telomere clock eventually ran down to zero. When disrupted in a causal experiment, the supposed master-regulator of aging and cancer did absolutely nothing for five generations.

You can also now "add time" to the telomere clocks. Nothing changes. Cancer rates don't increase, and in another misunderstood piece of simple math, they should theoretically increase by a huge amount. A linear increase in telomere length should set an exponentially larger limit to tumor growth: changing the Hayflick limit from forty-nine to fifty allows your cancer to be twice as large before it stops growing. Increasing the limit from fifty to one hundred allows your cancer to be 2^{50} times as large—1.1259×10^{15} (or 1.1 quadrillion) times the size. Tumor size should therefore be hypersensitive to telomere length. Yet when we increase telomere length, nothing happens. Tumor rates are effected by telomerase enzyme but not by telomere length. Instead, scientists don't see any effect, and we should conclude that telomeres don't limit tumor growth through gradual clock-like shortening.

Finally, we now know that telomeres are lengthened by telomerase in the body. Cell populations can double thousands of times in the body, even when they can only double one hundred times in the lab.[23] It is therefore not clear what role the Hayflick limit might have in limiting lifespan given that telomere length is so easily modified by our own bodies: whenever we need more cell divisions, we just "rewind" the cellular clock.

The link between progressive telomere shortening and aging is just another case of wishful thinking. Despite the clearly good intentions and immense intelligence of the proponents of the theory, it is wrong. Worse, there were clear signs that the theory was wrong from the outset. Simple math was either overlooked or left unmentioned. After thousands of academic papers and hundreds of millions of dollars in grant money, nobody had pointed out the trivial problems with the telomere theory of aging.

In the words of Professor Elizabeth Blackburn, who discovered the structure of telomeres and won the Nobel Prize for it, "You can always wish to see what you want to see."[24] In seeing the link between telomeres and aging, this risk seems to have been realized in dramatic fashion. Professor Blackburn, who seems a clear skeptic of the role of her work for aging, appears deeply aware of this risk and is worth quoting again: "we have to also be very sort of strict, because you know yourself. You're the most likely to be deluded, because you're the one who wants it to work, all right. So you also have to kind of play these two things off against each other . . . because in science you are trying to get at what is real. Right?"[25]

Right.

Unfortunately, Professor Blackburn would later shed her skepticism and write a typical lifestyle-advice column as a book, in which she advocated for ways to bring down your telomere length and halt aging.[26] Most of the advice was based on weak correlational observations or was unrelated to telomeres. I prefer to think that her heart was not in it because her earlier self-awareness gives me hope, and her later self-help book does not.

The telomere theory of aging should have become the story of a failure, used as a prompt for some deep soul-searching in aging biology. Instead, the link between telomeres and aging has been immediately and shamelessly rerouted toward new theories that bear no connection to the original hypothesis. A new and shiny distraction has been pasted over the wound.

One exercise in the power of denial, and the propensity for posthoc bullshit, comes from one of the most popular failures in redefining aging for the past twenty years. This is the heavily cited "hallmarks of aging" paper, which redefine the role of telomeres in aging. Hidden in the jargon is a complete abandonment of the original idea: "the concept of replication-induced senescence [aging caused by telomere shortening] as a general barrier requires refinement and reformulation. (Differences in telomere structure and function in mouse versus human cells have also complicated investigation of the roles of telomeres and telomerase in replicative senescence.)"[27] In other words, the original idea didn't work, and telomere shortening—the supposedly universal cause of biological aging—doesn't even extend as far as mice. Mice are the closest nonhuman organisms in which the idea could be tested in a lab, an evolutionary next-door neighbor to humans in the tree of life.

Senescent cells are another key to this shapeshifting magic trick. Senescent cells are slippery things to define. In the words of Norman Sharpless and Charles Sherr, while "the term 'cellular senescence' was originally used to define the state of irreversible proliferative arrest by cultured cells that had reached their replicative limit [the Hayflick limit or the 'cut-off telomere' state where a cell can no longer divide], it is now widely used to describe states of cell cycle arrest in different in vivo biological settings, including age-associated loss of regenerative capacity, tumor suppression, inflammation, wound healing and embryogenesis."[28]

In other words, at first there used to be one very clear definition of cell senescence, and now there are many, and most of these definitions seem to have nothing to do with one other. Sharpless and Sherr sharply continue that "a central problem in the senescence field is the lack of a uniform definition of cellular senescence, coupled with inconsistent application of biomarkers to identify and enumerate senescent cells in vivo."[29] For those who do not speak biochemist, this means that nobody can agree with what a senescent cell is or even say how to count them.*

So while telomere shortening used to cause aging in all species with telomeres by halting cancer growth, it now somehow causes aging through senescent cells—which nobody can measure, count, or clearly define.

Nobody explained why the previous idea failed after ten thousand or so scientific papers because of the shiny new distraction.

This has not stopped the telomere hype, which inevitably led to start-up companies.

All of the usual tricks played out. In 2019 I ran across the preprint for a new telomere paper contributed by a Nobel Prize winner† and a team from Harvard and Rutgers. They claimed to have increased mouse lifespan by 40 percent with a telomerase nasal spray.[30] After I pointed out their use of impossible sample sizes, the authors changed the number of mice in the experiment after the experiment was finished.[31]

This hiccup was only the beginning of the problems. As highlighted by Dr. Charles Brenner, the paper's peer reviewers—who are meant to act as independent gatekeepers of the scientific record—worked for the paper's authors and for BioViva, the company that owned the treatment and stood to profit from a positive review.[32] Dr. Brenner categorized BioViva as "a company that runs gene therapy trials abroad to skirt safety regulations and sells nonvalidated biological testing kits . . . bad actors."[33]

The paper's figures bore hallmarks of potential photoshop, with duplicated and rescaled panels inserted as (apparently) different mice: all detected by the amazing Dr. Elizabeth Bik.[34] This initially prompted the inaccurate nonexplanation that "it is required that the reviewers have expertise in what they are reviewing," before making a new and equally dicey explanation that "we have identified a duplicate FISH image" (taking credit for the work of Dr. Bik) and "this mistake of duplication happened during the image processing of multiple sections."[35] Another nonexplanation,

*Somehow, an avalanche of companies and foundations claim to remove or target senescent cells anyway, including the SENS Research Foundation started by the allegedly would-be paramour of young girls, Professor de Grey.

†The paper had been contributed by Professor George Church, a man who aims for a simultaneous cosplay of *Jurassic Park* and *Hair: The Musical* by resurrecting woolly mammoths. To help solve climate change. Maybe his plan works like putting on a winter coat, in summer, to make snow.

published in 2023 after a multiyear delay.[36] Unsurprisingly, Dr. Bik then highlighted that the replaced images had sections with different resolutions to the rest of the image, right over the top of the supposedly altered cells. This anomaly also went unanswered, the paper unretracted, as usual.

The creative types at BioViva were, however, only small operators. On a larger scale, telomere enthusiasm also gave us Geron, one of the most hyped companies in aging biology and the self-proclaimed "first biotech company to target aging." Incidentally, Geron holds another remarkable accolade: half of Geron employees are paid (and supposedly earn) over half a million dollars a year, the single highest median wage in California.[37] The CEO received about $5 million in 2023 despite only having about eighty employees. For this money, the company produced the only FDA-approved drug acting as a "telomerase inhibitor." Geron was launched with a flashy *Science* paper that claimed to extend the Hayflick limit to ninety population doublings for foreskin fibroblast cells.[38] I happen to know (for totally normal science reasons, I swear) that keeping foreskin fibroblasts in better cell culture conditions lets you reach one hundred doublings, without needing telomerase. Yet the authors were not worried by foreskin-related quibbles and started a company in 1998 that promised to deliver literal immortality within a decade.[39] The formula was simple: "Geron was hoping that by putting extra telomerase in cells, it would make them immortal, and thus reverse the aging process."[40]

The company initially bought a few Nobel laureates onto the board—including Professors Leonard Hayflick, Elizabeth Blackburn, and her co-Nobel laureate Carol Greider—but they are long gone. Blackburn left the board in 1994 for reasons unknown, and Greider left in 1996, which is strange given that Geron was about to make humans immortal.[41] It may be because Greider had the kind of integrity generally lacking in aging research. According to *Science*, Greider said that "the idea that shrinking telomeres play a role in organismal aging had been hyped in the media and by Geron's public relations division . . . [and] pointed out in her paper that longer telomeres do not correlate with greater longevity in animals." Greider then left because the board tried to stop her graduate student from publishing her own results.

In early clinical trials, the FDA placed a hold on Geron's telomerase drug for accumulated liver damage. Eventually, it let the company off again, and despite evaporating $635 million of Johnson & Johnson's money—the two companies have since parted ways—Geron is still in business. This is because the owners dropped the absurd premise that their drugs would make us immortal in 2011, after two decades of hyperbole, and did something else.

The new aims of Geron were slightly narrowed down, from curing all people of all diseases forever and making them immortal, to the rather more specific "treatment of transfusion dependent anemia in adult patients with lower risk myelodysplastic syndromes . . . who have failed to respond or have lost response to or are ineligible for erythropoiesis-stimulating agents."[42] Treating, but not curing, very specific types of cancer. This is the only time I can recall an antiaging company doing anything for the public good, but cancer drugs are not an aging cure.

It is worth recapping the history of the Hayflick limit: Weismann came up with an idea, and then because of the Nobel-Nazi Carrel, it had somehow become a universal dogma that cells were immortal. In the words of Hayflick, for "sixty years, from the time when cell culture began, in the late 1890s or 1900s, it was believed from virtually day one that cells put into a culture were inherently immortal," which was described recently as "mass self-delusion."[43] Then, because of Nobel laureate Hayflick, a new idea was developed: all the previous sixty years or so of thinking was wrong, and cells were not immortal in culture. Out of this grew the idea that this limit caused aging, and here we are another six decades later: able to modify the Hayflick limit, able to see that this intervention does nothing to the rate of aging, but somehow incapable of overturning the dogma once more.

Carrel debunked Weismann after a few decades, Hayflick debunked Carrel (and most of cell biology) after sixty years, and here we are sixty years later facing fundamental problems yet again. Instead of drawing the conclusion that history is repeating itself, and that we need to reexamine or throw out decades of work and thousands of papers, aging research is playing pretend.

Few have the stomach required to revisit the past hype. One of the few, it seems, was Hayflick.

After suffering four decades of bullshit promises about the impending cure of aging, the elderly Leonard Hayflick,* who discovered the Hayflick

---

*Anger that somehow did not prevent Hayflick from taking a paid position on the board of Geron. Perhaps it was the median salary of half a million a year that swayed him.

limit, saw things clearly when he said that "every five years, for the past forty years, there have been pronouncements made. . . . I'm still waiting. And I'm afraid I'm going to wait not only through my lifetime, but probably forever."[44]

Shortly after David Sinclair's mentor Guarente promised a new antiaging drug, Hayflick was absolutely and justifiably furious, stating "people like Guarente and [Professor Cynthia] Kenyon are not working in the field of aging at all."[45] He threw out a penetrating insight: "If they can show me the simplest way to prevent aging in their own automobiles, to have them live for a hundred years, then there will be some reason to buy into the biology argument. But they cannot do the simple thing, like keep their cars from aging for a twenty year period, to say nothing of biology."[46]

12 The Death of British Machines

Two demons burst into a house in India, one of them carrying a human corpse. While getting ready to eat the corpse, the demons begin to argue over who owns the dead body. As they start squabbling and fighting, they discover a traveler cowering in the corner in terror. The demons grab the traveler and demand he judge their dispute, and the terrified man reluctantly agrees.

Figuring he is doomed anyway, the man decides to tell the truth and supports the first demon as the corpse's rightful owner because it had carried the corpse into the house. Furious, the second demon then grabs the man and rips his arm off. The upset, corpse-carrying first demon responds in equal anger: acting instinctively, he rips an arm off the corpse and slaps it onto the living man like swapping a Barbie doll part. This escalates as the wronged demon tears off another arm, before the corpse-carrying demon undoes his work again and slaps on a replacement from the corpse. The traveler is still alive through this, but he now has two corpse arms and is presumably wetting himself. Yet the argument gets worse, and the demons alternately rip off pieces of his body in anger and slap on pieces of corpse in defense, until the man stands surrounded by the two angry demons and realizes he is *entirely made of the corpse*.

His original body lies in pieces on the ground, before the demons eat his old body in front of him and leave. In shock, and in a body made entirely of another person's corpse, the man staggers outside. His material self is gone, piece by piece, and his identity now resides in a new corpse-body that is not his own. He raves to himself, "Do I truly have a body now? Or do I have no body? If I think I have a body, what I have is completely a body of another; if I think I have no body, now I actually do have a body?"[1]

Predictably enough given the Indian setting of our story, the corpse man eventually stumbles upon some Buddhist monks as he desperately wanders the highways and asks them "Who am I, the *man* or the *corpse* or *nobody*?"[2] The monks mumble something unconvincing about how there is no "self" and leave the poor guy to his nightmarish existence.

This wonderful horror tale comes from the Mahāprajñāpāramitopadeśa (don't you know it?), which was fully translated into English by Jing Huang and Jonardon Ganeri just three years ago, and touches on a deeply philosophical point in which form is separated from identity by the replacement of organic parts.[3] I think the story is extraordinary both because it leaves the corpse man's questions tantalizingly unresolved—despite the Buddhist boondoggle of "there is no self"—and because it comes down to us from around 300–450 BC.

This is centuries before Plutarch[4] alluded to the ship of Theseus, a similar conundrum in "Western" thought.* This better-known idea, which sadly lacks corpse-eating demons and monks, talks about a problem that has consumed philosophers for two millennia. If a complex form is replaced piece by piece—in Plutarch's example the form is a ship—is a new identity created or does the old identity survive?

While this may be a fun piece of philosophy, a reason I enjoy this problem is that it also contributes to one of the most popular, and indeed now almost all-pervasive philosophical ideas that we have: the picture of ourselves as machines.

Previous thinkers had compared organisms to physical machines using varying degrees of analogy, but this usually involved comparing humans to a sort of puppet moved by God, the Gods, or even the Fates rather than physical laws.[5] This role for Gods and mystic forces was quietly omitted by

*Certain kinds of people seem to love the boringly uniform pretend entity called "the West," a concept that swallows complexity. The preservation of early philosophy routinely rests on the translation and preservation work of the Baghdad court: just one of the many guarantors and generators of "Western" knowledge. In this case, however, it was the Byzantines—which resemble modern Western countries about as much as I resemble an Etruscan shrew—who preserved Plutarch. The Byzantines sent Plutarch's writings back to Italy around the eleventh century, where he remained largely unread until the early Renaissance. His *Parallel Lives* containing the entry on Theseus was only printed in Venice as late as 1517, the first time the idea was popularized west of Constantinople since the fall of Rome a thousand years before.

the early Renaissance anatomist Vesalius, who didn't mention the supernatural like his predecessors, but an active exclusion of the Gods took another eighty years.[6]

Our modern understanding of organisms as a machine is largely inherited from the philosopher-mathematician René Descartes, who had the audacity to picture reality as a radically and fully mechanical universe governed only by the laws of physics—until the Inquisition and the Pope got involved.

Descartes initially considered everything to be determined by physical laws, with no exception for humans and "at first placed in [humanity] no rational Soul."[7] This was a staggeringly original and, for the time, very dangerous idea: one that Descartes rapidly dropped when the Pope declared Galileo a heretic and a criminal in 1633. Galileo was prosecuted merely for treating the solar system as if it were governed by physical and mechanical laws, while Descartes was proposing that all the universe and life itself was not the realm of God but of physics alone. Descartes revealed his resulting panic about the prosecution of Galileo to his friend Mersenne: "I was so astonished at this that I almost decided to burn all my papers. . . . I could not imagine that he [Galileo]—an Italian and, as I understand, in the good graces of the Pope—could have been made a criminal."[8]

So, it is courageous that Descartes did not abandon his idea of a clockwork universe entirely and instead made a crafty workaround: he placed the soul in a little hidey-hole, the human pineal gland, which he said contained the seat of all reason set apart from complex animal "reaction."* Although the timing of his soul insertion is very suspicious, it does not seem that Descartes was necessarily fibbing. His much later correspondence with a brilliant and very rich† patron, princess Elizabeth of Bohemia, seems to indicate that Descartes was perhaps sincerely devoted to his "soul in the

---

*This strange exceptionalism for "man" led to a rather funny-not-funny debate by the Calvinists over where to draw the line, including whether women were included in the special soul-carrying club or were, like animals, just clockwork automatons with complex responses. Some people lumped in animals with humans—following the surprisingly old Christian argument that dogs do indeed go to heaven—and others excluded all animals from heaven and tried to add women to the crowd of the soulless. The moral of the story is don't time travel to the past, even the recent past, because it is usually terrible.

†My colleagues at rich academic institutions have certainly revealed how flexible (absent) their morals become when a rich patron arrives, so who knows what Descartes really made of things in private.

pineal gland" solution.[9] The rest of the universe, including animals and our own bodies, remained clockwork driven and obeyed the (largely undiscovered) laws of physics.

Publishing anonymously at first, Descartes catapulted his ideas into the Renaissance and somehow got away with it. The Pope stayed quiet,* and the Inquisition didn't even try to burn him alive, as they had done to Bruno and attempted to do to Galileo—who was ironically saved by his mate the Pope. Descartes's bold idea of life as a clockwork affair driven by physical-mechanical forces remained intact. Humanity was just automatons pushed around by physics, with no role at all for horoscopes or mystical forces, and a suspiciously tiny gland as a placeholder for the Almighty.[10]

We were machines.

Scientists and the public have absorbed this idea so completely that it is difficult to grasp how radical the idea was in 1637. There was some idea of anatomy as a physical system, but this anatomical thinking was also a wild new concept at the very forefront of knowledge.[11] Few people in the 1630s would have ever seen a complex working machine, with most being simple milling devices that would seem farcically separated from the functioning of organic life. Even clocks, probably the most complicated machine most people would ever encounter, were clunky, mystical status symbols of the church, housed in cathedrals with the holy relics. Greater rarities† like wind-up automatons were so strange, expensive, and exciting that they were produced in just a single city, Augsburg, and were owned by kings as status symbols. The Holy Roman Empire used wind-up ships to seal peace treaties, with great effect because they gave their owners massive soft-power clout.[12] Their successors the Habsburgs even paid the Ottoman Empire's ransom—when the fate of all Eastern Europe was at stake—by sending them clockwork trinkets like the "Sultan on Horseback" instead of piles of gold florins.[13]

The idea that we, and all of life itself, are more elaborate versions of "automatic" devices was therefore utterly radical in scope. Even more

*Perhaps he was thinking of 1 Corinthians (NIV), where the physical or natural body is separated from the spiritual body. The bible is mute, as far as I know, on why the physical body is "perishable" in this scenario—the church seems to just say you have an "allotted" time on earth.

†About a hundred European-made automatons existed when Descartes was writing.

amazing was that Descartes and his contemporaries also understood the challenges such an idea held for aging, identity, and mortality.

One challenge was a resurrected version of the ship of Theseus and the Mahāprajñāpāramitopadeśa demon story problem: if I am a mechanical device and my body can repair itself, how do I retain my identity while my parts are slowly replaced? If my identity is purely physical, not metaphysical,* is my identity slowly eroding or dying whenever my body is repaired? Mind-body duality, the split between a soul and the body, only displaced this problem: like the monks in the Indian tale, it did not really answer the question, so much as shove it off to the side.

Even greater, however, was the question of aging. What made your body decay? Even if you accept it, Descartes's type of mind-body duality, where an imperishable soul (or mind) lives in an imperishable body, does not tell you *why* the body is perishable and slowly decays with age.

Treating the body as a machine allowed a fascinating new line of reasoning in this direction. The few people who did understand simple machines knew that they broke down and, vitally, that older machines seemed more likely to break than new ones. They also knew, thanks to philosophical problems like the ship of Theseus, that machines could be repaired indefinitely.

Transplanting this reasoning into biology introduced all kinds of incredible possibilities and problems. All good gardeners know you can propagate entire new trees (and grapevines) from cuttings, and philosophical types had recognized that plants could therefore completely repair and replace themselves indefinitely. If organisms were repairable machines, this meant that complete replacement was somehow possible. Early on, there had been a division drawn between plants and animals to avoid this issue: plants were capable of indefinite repair, for unknown reasons, and animals were (supposedly) not.

---

*Descartes's hidden soul was a way to skirt this problem: if the identity of rational beings was immutable and hidden in the pineal gland, slow parts replacement does not affect identity. But this answer didn't really satisfy anyone and obviously just transplants the same question onto anything without the required soul-gland. What happens to the identity of my (apparently soulless) pet if his parts are slowly replaced? Is my fish Radagast permanent or impermanent? Does he lose his identity, little by little, every time a fish scale is replaced?

This division was blown apart in 1744 when a scientist named Abraham Trembley began investigating pondlife with his newly invented microscope. Trembley found that "animalcules" called *Hydra*—a little glassy ball of tentacles that seizes and eats other pondlife—could also indefinitely replace themselves from cuttings.[14] You could slice *Hydra* in half, and they would completely replace the missing half. Slice a head off, and like the Hydra in the Greek myth, another head would grow in its place. You could do it again, and again, and again, like growing a grapevine or an apple tree from a cutting. The *Hydra* was capable of a perfect repair job, every time.

As a result, indefinite, near-perfect repair was revealed to be possible in animals: look at *Hydra*, planarian flatworms,* lizards regrowing tails (to a less perfect extent), and axolotls regrowing limbs. After Trembley, there was no clear division between animals and plants. The growth of microscopes compounded this picture when it also became clear that while we cannot regrow limbs, we were constantly regrowing cells and we had replaceable parts.

All of which posed enormous conundrums if your body was a machine.[15]

To illustrate why, and help us solve some more modern problems, let's bring back our two demons from the mythical Buddhist story. First, we should give them names. I could borrow two wonderful names from the *Ramayana* and call the demon that rips off body parts "Māra" and the one who places them back "Ravana," but to make things simpler, let's call them Breaker and Builder. Our demons will still tear off pieces (Breaker) or replace them with a new part (Builder), but this time let's shrink them down to microscopic size and allow them to them work as fast or as slow as we want.

If Breaker tears off microscopic pieces of your body and Builder does nothing about it, you will eventually stop functioning and die. Because we made Breaker microscopically small, this process may even be gradual, like the aging process.

But what if we allow Builder to put pieces back on?

Well, if we assume that Builder does a perfect repair job like the demon in the story, the result will depend on speed. If Breaker is faster than Builder, Builder will just slow down the inevitable. You will "age," in that the systemic function of your body will be composed of more and more broken

---

*You can cut a planarian flatworm in half, and it will repair itself indefinitely, like the broom in *Fantasia*.

parts until you cease to function entirely (figure 12.1). That is, the slower the repair rate, the faster you fall apart.

However, if Builder is just as fast or faster than Breaker, you will always get your microscopically broken body parts repaired after a little delay, and your mechanical bits and pieces will soon reach a steady state. That is, once the two demons settle down to a rhythm, things won't get worse over time. At any one point in time, a small part of you will be newly broken, but that broken part will be taken out and replaced by Builder before the damage builds up too much (figure 12.1). If Builder keeps up with Breaker, aging only happens for a short period of time before your body reaches a steady state of slight disrepair.

The same "aging" process may also apply if Builder does an imperfect job of repair. Badly repaired parts may gradually accumulate in your body, "aging" you, as more and more of your body is composed of imperfectly repaired parts. However, the window in which this is achievable is limited to a narrower range of replacement rates. If the process of replacing good parts with less reliable ones continues, eventually you become entirely composed of imperfectly repaired parts and aging stops: your completely replaced body is bad, but it doesn't get any *worse* as time goes on.

Less obviously, there are also situations where imperfections in repair affect how well Builder demons work: that is, an imperfectly replaced part lowers the overall quality of repair, which lowers the quality of parts replacement further, in a feedback loop that eventually destroys the system. This theory is, wonderfully, called the "error catastrophe" theory of aging when applied to our systems of genetic repair.[16]

This all means that aging depends on balancing the rates of damage and repair.

These theoretical models allow aging to occur in repairable systems, under certain circumstances, and somewhat solve the problems highlighted by Trembley. Under certain circumstances—like if Builder is slower than Breaker—aging will theoretically happen in a repairable machine.[17] Our best model of this process is vehicles, which slowly decay in function with age despite being repairable.[18]

Unfortunately, these ideas really are just a model. Back in WWII, engineers used these general ideas to develop what we call "reliability theory" and simulate how long machines will survive. I am going to avoid an overview because otherwise we will go down a rabbit hole of parallel lightbulb

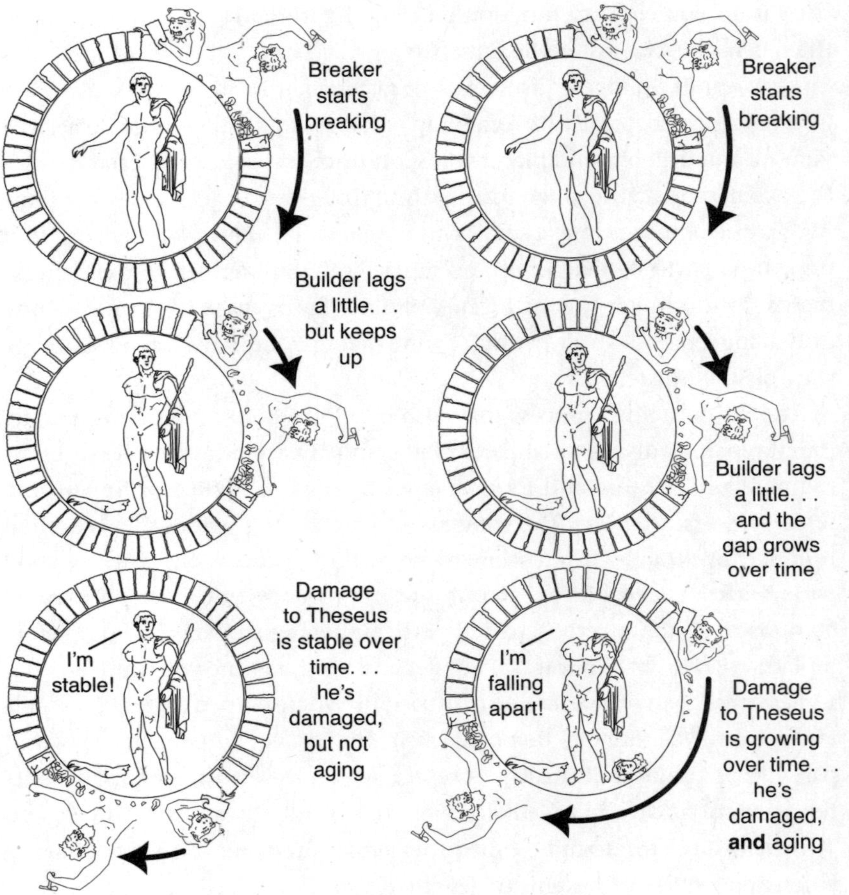

Figure 12.1
Aging pictured as a balance between damage and repair. If we picture your body as consisting of repairable units (the bricks on the outside), the rate of decay and even the presence of aging depends on the balance between two factors, here drawn as a "breaker" demon (shattering bricks with a hammer) and a "builder" demon (putting them back together). If Builder is faster than Breaker, we descend the left-hand cartoons: damage to your body is repaired as rapidly as it occurs, with some finite delay, and your body does not progressively decay over time. Some part of you is always damaged, but the damage does not accumulate with age. Otherwise, if Builder is slower than Breaker (or gets slower with age), the gap between damage and repair grows over time, and your body progressively decays to cause aging (right-hand cartoons). The second idea is a foundational picture of repair processes in the body, where damage (supposedly) accumulates inevitably over time. A similar theory plays out if repairs are imperfect and errors in repairing the body accumulate over time—though after fifty years, the evidence for this is still theoretical. Source: Author.

circuits and risk death by boredom. Instead, I will just say there are many theoretical models of mechanical repair that predict aging, based on theoretical damage and repair patterns in machines.

The engineers have largely moved on, developing theoretical models where aging does not happen or where machine survival curves take an arbitrary shape. Biologists and aging scientists, however, continue to use the old models of aging despite knowing that there are many species capable of perfect repair, because they cling to the idea that physical damage theoretically causes aging in mechanical systems. With extremely rare exceptions, biologists and aging researchers are not observing what happens to complex machines as they stick to old models of machine survival.[19] This is dangerous because so much depends on these old ideas and on a theoretical understanding of how machines age, inherited from a seventh-century polymath spy who had seen a clock.[20]

To know if damage-based theories of aging work, we need to observe the death of machines. We need to see Breaker and Builder in action because in the real world, math-based models often get their teeth kicked in.

This hole in our philosophical picture can be plugged up in the modern era, not because we can suddenly outsmart Descartes, Nagasena, or Heraclitus but because we are building billions of machines and releasing them into the wild. This allows us to move beyond theoretical musings to hard-nosed observation.

During 2021–2023, I convinced the UK government to send me information on every single British road-registered car, motorbike, three-wheeler, ambulance, van, light truck, and moped in the country: all of which underwent a roadworthy inspection, a kind of annual census of machines, with repair records and mileage added on as a bonus. More than ten billion recorded parts failures were observed at these roadworthy inspections, encoded in British style by thousands of uniquely numbered mechanical failures. To make the data even more complete, the Brits kept certificates of export and destruction for the entire country, reporting the exact dates that each vehicle crashed, died, or emigrated from the Soggy Isles of Britain. In the end, I assembled data on the life, migration, and death of sixty-five million machines and enjoyed the immense pleasure of analyzing them.[21]

This rolling annual census of machines, and their broken-down parts, was full of practical and philosophical surprises.

For example, the conundrum of the ship of Theseus—of whether physical parts replacement erodes or destroys identity—was solved in a delightfully pragmatic way. I can replace every part in my little car, piece by piece, indefinitely: what defines its identity is not its individual or collective parts, nor even its number plates—which are transferrable to a new vehicle. The identity of my vehicle is not physical at all: it is defined by the International Organization for Standardization's Vehicle Identification Number (VIN), a string of letters and numbers that uniquely defines my vehicle worldwide, regardless of how many parts are replaced, until its identity is administratively destroyed.[22] Who knew bureaucrats and engineers could solve philosophy problems.

Another philosophical idea that fell over, in a more meaningful and less amusingly pragmatic way, was the long-standing claim of equivalence between human and machine survival patterns.[23]

In chapter 2 we encountered the human survival curve, which has a characteristic post-birth "bathtub" shape that falls from the peak of high mortality risk at birth down to an adolescent low, before rising roughly exponentially with age, with the risk of death doubling every eight years or so until death (see figure 2.2). At the end of this process is the highly contentious flattening out of mortality curves at advanced ages.

Vehicles have long been observed to, supposedly, follow the same general pattern, including the questionable flattening out part. This shared pattern has supported the idea that humans age and "wear out" in the same way as mechanical devices.[24] However, these observations were based on just a handful of car populations, observed only until their fifteenth to seventeenth year or so.[25] In contrast, my British data spanned over six thousand common vehicles of all ages and types, each population of which was observed across seventeen years from their starting age, which was often quite old.

What was the pattern of machine longevity?

At first, vehicles generally, but not always, behave roughly like humans.[26] The odds of death rise exponentially with age, doubling every year or two depending on the make and model of the vehicle (figure 12.2). This was only a trend, and there were many oddities like British black cabs* and

*Black cabs had an almost flat mortality rate, rather than an exponential increase in mortality, and almost never died from mechanical failure. Instead, they were forced to die at age fifteen through an arbitrary government decree, after a lifetime of negligible aging rates.

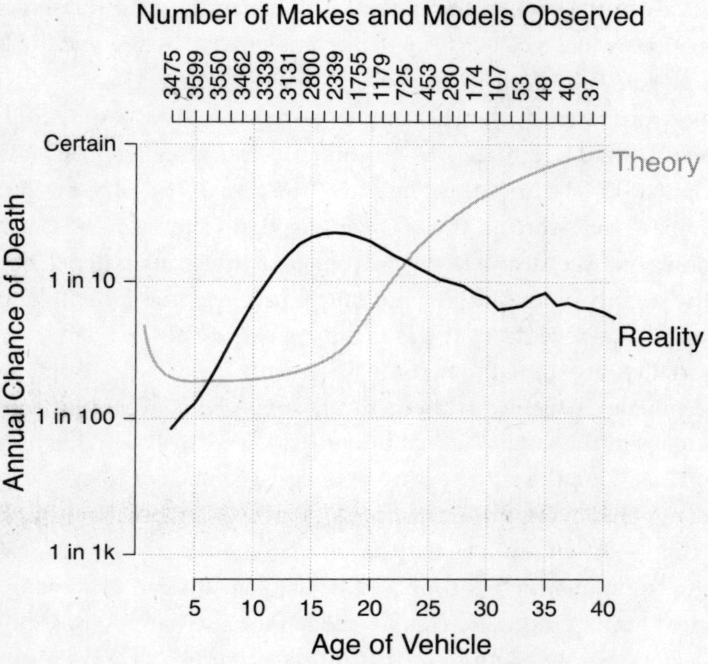

Figure 12.2

The odd patterns of mechanical mortality. Machines are our intellectual model of aging in organisms and are supposed to have mortality patterns that resemble post-birth humans (the gray "theory" curve), with high "infant" mortality after manufacturing, followed by an accelerating mortality rate until machine death (see chapter 2). In contrast, the survival rates of tens of millions of vehicles (excluding exported and crashed vehicles) reveal how machines do not behave at all like our theories expect: at first, mortality rates usually accelerate (get worse) with age, before inverting to "age in reverse" by becoming less likely to die with age after roughly their seventeenth to eighteenth year (black "reality" curve). However, this is just a general pattern. Motorcycles, black cabs, and vans all behave very differently, none of which fits our biomechanical theories of aging. Similar patterns happen for mortality rates per mile driven. Source: Author.

mopeds. Even the general trend got weird, however, when I observed past the previously observed fifteen- to seventeen-year mark and looked at advanced ages.[27]

In previous research, vehicles on the mortality rate curve did flatten into a plateau toward age fifteen to seventeen before they stopped observing the populations.[28] But is this "late life" for a car? The oldest vehicles in my database were born in the 1890s and still alive in 2021, so it is highly debatable, but the distinction was rendered irrelevant when I observed even the slightly older cars. The mortality curve was not flattening out as it approached age seventeen. It was turning a corner. Beyond age seventeen, the mortality rate started to become lower with age.

The vehicles, which were theoretically supposed to accumulate irreversible damage in the same way that humans do, were instead aging in reverse (figure 12.2).[29] For the entire period from age seventeen to forty, the mortality rate of vehicles fell linearly with age. The vehicles became less likely to die as time went on, even as they accumulated more rust and damage and mileage. None of which is supposed to happen. Instead of looking like a "bathtub" human mortality curve, vehicle mortality curves look more like an anteater: sharply rising from the nose, flattening at the shoulders, and slowly and gracefully curving down across the back to a long, swooping, and increasingly fuzzy tail (figure 12.2).

The same pattern emerged when I measured age in miles instead of years. Initially, the vehicles would be more likely to die with every additional mile until they reached about 150,000 miles. Then, the mortality rates flattened and inverted to fall with each mile in a long and steady reduction in the mortality rate. By the time vehicles had driven 400,000 miles, their mortality rate had fallen by so much that they matched vehicles that had been driven only 60,000 miles. It was weird but quite consistent across models and makes.[30]

These patterns do not seem to be caused by age-coding errors. As well as number plates, vehicles carry standardized unique logbooks and have internationally standardized codes welded to their engines and multiple places on their bodies at birth (the VIN).[31] Unlike humans, vehicles are also usually obviously identifiable by year of birth, in a way that becomes more obvious as the error size increases. It is conceivable that your mechanic could mistake a 1978 Ford Mustang for a 1980 model but unlikely they cannot tell the 1978 from the 2021 model. Finally, vehicle data are independently reported. So to make an age error, every mechanic performing a roadworthy inspection on the vehicle would have to consistently report the wrong make, model, odometer, VIN, and number plate in a logbook somehow reissued to the incorrect year, in a way that matches all the other mechanics who ever inspected the vehicle. All of which is to say, I don't think it is a mistake.

The long reversal of mortality rates with age also had a strange parallel when I observed the odds of parts failure instead of the odds of death. Consider a 2002 Ford Fiesta, one of the most common vehicles in the database. After driving 20,000 miles, about 10 percent of 2002 Fiestas failed their roadworthy inspection due to major problems—a major parts failure. Add 40,000 extra miles, and this number doubles to a 20 percent failure rate. Add another 40,000 miles, to 100,000 miles total, and the failure rate *almost* increases by 10 percent again, to just below 30 percent. To be clear, this is a thoroughly decent level of reliability. However, the relationship falls apart once the vehicle gets too "old." Put more miles on the odometer, and the odds of failing an inspection flatten out: the mechanical rate of failure does not get any worse as you drive farther and farther. In the jump from 100,000 to 200,000 miles, the odds of failing a roadworthy inspection for mechanical reasons only increases by 7 or 8 percent. By 300,000 miles, the odds of failing an inspection don't increase at all, and they may even decline slightly. A linear increase of the amount of "wear and tear" no longer causes increasing odds of mechanical failure.

Correcting for past rates of mechanical failure didn't change these striking patterns and neither did correcting for owner's average income, the road network, the location of the driving, or basically anything else. I even fit machine learning (AI) models to the data, adding every predictor I could to predict survival, and the mortality rates still fell with age. In aggregate, across all vehicle types, there is again a reduction in parts failure rates over time. As a vehicle gets older, it becomes very slightly more reliable at very old ages.[32]

The general pattern of vehicle survival is highly unusual, and at older ages, the machines did something that should have been theoretically impossible under our existing models of aging. They became progressively less likely to die with age—regardless of whether age was measured in years or miles driven (figure 12.2).

Machines can age in reverse.

I am going to take a brief side excursion from the main narrative for a few paragraphs because the vehicle survival data were so bloody interesting. These data allowed me to test all kinds of fascinating stuff—like what vehicles are the most reliable and which parts of the United Kingdom contribute the most to climate change.[33]

So what are the most reliable used cars, vans, and motorcycles?

First let's just admit that this *should be* common knowledge. Vehicles are expensive, and everyone who buys a car, van, or motorcycle wants to know which vehicles are the most reliable. Stupidly, we have an elaborate and independent system for measuring which vehicles are "alive" or registered and a system for regularly checking if vehicles are working well mechanically, respectively, by registering the vehicle and by sending it in for government-mandated roadworthy inspections. These things are paid for by our taxes, registration fees, and our fees at the mechanics in virtually every country on earth. Yet our governments don't tell us what vehicles last the longest and pass the test most often.

I took offense at this oversight and made data available at DriveDictionary.com so you can see for yourself, for all six-thousand-plus used models in the database, what vehicles had the best reliability and longevity.[34]

In short, the longest-lasting vehicles are those used as taxis and for businesses. For UK cars made before 2019, the most mechanically reliable common vehicles are the 2014–2018 Toyota Prius and the 2017–2018 Mercedes E-class, both commonly used as taxi cabs. The most mechanically reliable vans seem to be the 2014–2018 Mercedes-Benz Sprinter van and the 2015–2018 Iveco Daily van. The most reliable motorcycle is the 2018 BMW R1200, and the most reliable moped is the 2017 Honda NSC. Although let's be real, nobody buys a motorcycle for reliability—it's to try and look cool and have fun—and mopeds seem to exist to propel people on gig-economy poverty wages through traffic while delivering takeaway.

The tables I generated on car survival and reliability certainly could help settle a lot of tedious arguments and satisfy obsessions of the mechanically minded over "what car is better." Curiously, however, reliability was not just about mechanical quality. Manufacturers often sell the same vehicle under different brands (at different prices) within one country. Mechanically, a Volkswagen Golf is both an Audi A3 and a Skoda Octavia: they come off the same factory floor, with the same engine, and they just change the cosmetics. In some models, you can pry off the Skoda badge from the engine and find a VW badge underneath. Despite this mechanical cloning, split-badged models display very different reliability rates and longevity. They all die at very different rates, per year and per mile, all of which is kind of fun and interesting.

After this analysis, I moved on to climate change: the specter that will haunt us all, right up until our grandchildren, the poor, and the polar bears get what the oil executives deserve. With my colleagues, we discovered that the United Kingdom was exporting high-polluting cars to lower-income countries while, cynically and incongruously, sending the clean versions to the scrapyard.[35] Most exporting countries are compounding this problem by cutting out and "recycling" catalytic converters—which reduce toxic air pollution by 90 percent—before they export the vehicles. A triumph of two-faced rich-country cynicism. Other air pollution reduction devices are also routinely (and legally) stripped. Even the airbags are removed for parts, which the poor-country buyers may only discover when their head strikes the steering column, because a $4 net profit for a scrap-metal merchant in Preston is somehow more important than a human life in Kenya.[36] All of which means that poor people get vehicles with no emissions reduction devices whether they like it or not (they don't), while rich countries keep pretending to fight climate change, because we don't feel like closing an easily closed export loophole.[37] Even worse, the non-CO_2 air pollution from these vehicles also kills millions of people a year, helped along because we allow dirty cars to be exported for a tiny net profit.[38]

This rich-person cynicism played out on the British scale as well. Rich people drive the most, produce the most CO_2, and put out the most carbon monoxide per mile. The richest 10 percent of people generated about double the emissions per mile, compared to the middle or working class, and they drove far more often. At the same time that I made this analysis, the richest boroughs of London were busy complaining that they could not afford electric cars because it was too expensive. They did this while driving luxury cars that, if you sold them second-hand, would leave you enough money to buy three to five new electric cars. Trading in Rishi Sunak's fuel-guzzling Jaguar, which he drove much farther than his "I understand the poors" hatchback, would have allowed him to buy about six "unaffordable" electric cars.

Then, with this study of double-standards complete, I turned back to studying the death of machines.

Machines can age in reverse, becoming less likely to die with increasing age and mileage.

The possibility raised by Descartes, elaborated by many others, and shaped into its modern form under reliability theory, was this: humans break down and wear out like machines.[39] Yet academics* don't really have a firm understanding of machines in the wild, and it shows. Machine aging is no more inevitable than biological aging. There is a general pattern that older age is linked to higher mortality, but there are so many abundant exceptions that it is clear we are missing something entirely.

We can make a guess at what that "something" is if we bring our demons back into the narrative.

The Breaker demon in a vehicle is what we would think of as wear and tear—the mechanical wearing out or degradation of parts, like your tires going bald or your brake drums rusting out. This should be roughly constant over a fixed distance or a fixed amount of time unless there are truly massive differences in how or where a vehicle is driven. We can account for the "where" as most driving is done within about ten kilometers (or six miles) of your front door. However, the antiaging patterns remain after you adjust for location with a fancy-doodle machine learning or AI model. This just leaves *how* you drive, which could change the rate at which parts wear out. Yet this would mean that physical parts would also fail their roadworthy inspection more often with age: instead, they fail less often once you get past the "hump" age of seventeen.

So it's probably not Breaker—changing patterns of wear and tear—but what about Builder?

If part of my car is damaged, I can replace it by slapping on the scrounged part from a car corpse. I can change a tire, and (in principle†) any other part of the car. A mechanic can do it even better. What does this mean? Well,

---

*Scholars at public institutions, at least. I suspect engineers and mechanics know exactly how much vehicle reliability depends on the owner as well as the car.

†My crowning automotive achievement came in university when I was utterly skint, and my ancient Mazda 626 station wagon was broken into for some loose change and a pair of sunglasses. The thief smashed the window and wedged a pair of scissors into the ignition barrel before snapping them off at the blade, so I hotwired the thing and drove around junkyards until I found a new driver's side door. I fixed the door and electrics but never the ignition, so I just left the wires hanging and touched them together to get her started. Not the greatest Builder demon effort, hence the "in principle."

like the rate of damage, the rate and accuracy of repairs is not necessarily constant. As a vehicle gets older, the rate at which a Builder demon (or mechanic) does repairs can change. It *does* change: sometimes randomly, sometimes systematically. Repairing a vehicle comes about through a combination of personal and economic decisions. I *can* repair a vehicle forever if I can afford it. I *will* repair a vehicle forever but only if I both feel like it and can afford the repairs.

This has an unusual practical outcome. A Builder demon, by which I mean a mechanic, will replace a vehicle with fresh corpse parts as long as you pay them. The upshot is that vehicle longevity is ultimately determined by economic factors, almost independently of the vehicle's mechanical reliability. When a vehicle part breaks, you decide on whether you replace the vehicle or scrap it. Machine survival is then a personal and economic decision, not an inevitable outcome of physical damage.

Some evidence of this comes from economists themselves. Vehicles are mainly produced in a small number of places but exported worldwide: for example, the Mini Cooper is currently only assembled in Oxford.[40] However, as much as 80 percent of the variance in vehicle lifespan may be determined by the economy of the country where the vehicle is sold, not the make and model.[41] More specifically, how long a vehicle lives seems to be hugely affected by the ratio between the cost of a new vehicle and the wages of the mechanic doing the repairs.[42]

A long time ago, in a footnote of my obscure doctoral thesis, I decided that an effect of this process was that "decreasing costs of system repair relative to system replacement increases system longevity indefinitely. No single 1965 Ferrari GTO has ever undergone catastrophic failure, not because it is a reliable system, but because the cost of a Ferrari GTO has grown exponentially for the past 50 years. Ferrari GTOs are effectively immortal, because they are two orders of magnitude cheaper to repair than to replace."[43] In the nine years since, this idea has held up, with zero GTOs dying and their sales price doubling to over $50 million.

An extreme case of this occurs in Cuba, where 1950s and 1960s vehicles drive the streets in near-perfect condition. Tourists seem to think this is whimsy but instead it is the result of shockingly low wages and the historical near-impossibility and astronomical cost of importing a vehicle. A haircut, a universal indicator of labor costs, is about $1 in Cuba. Cuban classified ads for 2024 reveal that a two-year-old Toyota 4WD costs $95,000 or more: about 7,916 years' worth of haircuts. In the United States this car cost about $25,000 to $30,000, or 80–90 years' worth of haircuts at US prices. Cuban vehicles cost a fortune to replace but are repairable for almost nothing. The outcome is that Cuban cars have extraordinary longevity because the invisible hand of economics extends their lives indefinitely.*

This effect also plays out in the United States. Vehicles appeared more reliable in the late 1990s than in the 1960s, but a lot of that was because of "an erosion of repair-and-replacement parts prices" so that "essentially none of the improvement in car longevity is attributable to improvements in the inherent durability of cars."[44] After labor prices crashed at the end of WWII, vehicles lasted roughly three years longer than in the 1970s, and the relatively moderate lifespans of vehicles now reflects the rising costs of labor.

The other part of the mystery is personal preference—which often determines how the invisible hand of economics behaves by changing how much we are willing to pay to keep a vehicle running. With a new vehicle, this is usually quite pragmatic. We make a decision at the mechanics: pay for a new clutch or cut our losses and buy a replacement car. With an old vehicle, however, repair decisions can be emotional. In an industry-rag survey of classic car owners, who are *overwhelmingly* men, 62 percent said that their classic car was the most important thing in their life.[45]

I cannot know for certain, but these two forces—money and emotion—are my best guess at the antiaging of vehicles at older ages. As vehicles become older and higher mileage, people form (or are more likely to have had) an emotional attachment to them. This in turn means that maintaining and repairing the machine becomes *more likely* with age, reducing the risk of mortality over time. In conjunction with this process, older vehicles also appreciate in value at older ages. The economic payoff of repair becomes higher and higher: the extreme case of this is the $50 million Ferrari, which you would hardly send to the scrapyard because you didn't want to buy a new spark plug.

At younger ages, survival may also have a lot more to do with economics. I don't know that it is a coincidence that vehicle sales values follow the inverse of vehicle survival rates: they fall exponentially until about age seventeen before they slowly rise again as the vehicle becomes a "classic"

* If the embargo is ever lifted, or cheap imports flow freely into Cuba, the old American Cadillacs will die very rapidly.

and appreciates in value. This process gradually increases the cost benefit of paying a Builder demon to perform replacements, with greater accuracy and care, and may lead to a progressively falling risk of parts failure or death. Antiaging yourself through rising asset values.

Mechanical systems, which we thought we understood, can be very strange indeed when parts are indefinitely replaceable. Sixty-five million machines certainly behaved strangely and, in many cases, become more likely to survive with age even as wear and tear progressively increases.

This seems especially confusing if, as I suspect but do not know, machine longevity is being modified by things like economic costs, price depreciation, and appreciation curves, and even the emotional trajectories of old British men. Even without such strange modifiers, however, very little of the observed patterns of machine death make sense.

Scientists have based a huge amount of aging research, and marketable antiaging interventions, on the premise that we must prevent the accumulation of physical damage. We are told we must prevent "damage" or wear and tear—from stress, bad skincare, too few antioxidants, or too much bacon—to prevent aging. We have also generated a lot of theoretical thinking on the idea that aging is caused by the body progressively accumulating damage over time.

Yet if the longevity of machines like cars is not constrained by wear and tear, where does that leave our model of organisms as machines? If we didn't even manage to understand the long-standing model of our bodies, in cars, what does that mean about everything else? What does it mean about us?

We need to take a much closer look at how parts replacement happens in humans.

13 Oysters à la Cyanide

Dead humans follow the boring kind of chemical reaction: the type that settles down to a pleasing yellow, a disgusting foam, or a corrosive gas. Because your chemistry generally follows the ordinary, non-oscillatory, everyday kind of chemistry, your rate of decay is quite predictable: your molecules decay at a fixed rate, much like a radioactive half-life, determined by something called first-order kinetics and the law of mass action.[1]

Exactly what this means can be explained by a thought experiment where I poison you (my dear reader) with a huge dose of cyanide. So, let me kill you, and then we can talk about your rate of decay.

I specifically chose cyanide for my thought experiment because of the Krebs cycle, which is the cellular engine your cells use to generate energy. Cyanide stops the Krebs cycle immediately. It is like throwing a stick into your cellular bicycle spokes, without really breaking the spokes themselves. Everything stops dead, but the machinery itself isn't shattered. Everything is largely fine—perhaps a little bent in places—and the gears won't turn anymore.

When I kill you with cyanide, several things happen in rapid succession. Your heart stops beating, which halts the oxygen supply to your cells. Your blood begins to clot, and your body begins to cool down to the ambient temperature. For reasons I will make clear, I don't want any of this. Therefore, in my homicidal thought experiment, imagine I hooked up your soon-to-be corpse to a medical device called an ECMO that pumps you full of oxygenated body-temperature blood and removes the used-up blood at the normal rate. I also keep you warmed up to body temperature. You stay warm and richly oxygenated, like you would in life.

Why on earth would I perform this grisly and elaborate thought experiment? What on earth would I gain from killing you, without damaging you (much), and while keeping you at ambient "living-style" levels of temperature and oxygenation? Well, I am trying to achieve a scientific ideal called ceteris paribus: to perform a good experiment, you should keep all unimportant things in a constant (enough) state. If you conduct an experiment with "all other things being equal," or ceteris paribus in Latin, you can be (reasonably) sure that the result of your experiment happened because you modified just that one thing and not because some other thing changed by accident.

When I kill you with cyanide, I'm trying to shut off all cellular repair mechanisms without really breaking anything else so that I can then watch how fast your physical machinery decays when it is still oxygen rich and body temperature. Basically, cyanide stops our little Builder demon working immediately, without dramatically changing the rate that Breaker is, well, breaking stuff.*

So if we stop Builder from working and keep everything else about equal, how fast do you rot out? The answer is *very fast*.

You can even conduct this kind of experiment yourself, if you wish, on a smaller non-homicidal scale. Take a warm-water bath and an oxygenation device or even better a tropical fish tank (which supplies both warmth and aeration). Dust some oysters or mussels with slug bait, and drop them in your tank.† Oysters don't have circulatory systems to speak of, so no blood-pumping ECMO machine is needed. Just wait and observe.

A corpse in a body-temperature, wet, oxygen-rich environment will take days or even hours to disintegrate.

Why is this very familiar concept, that corpses degrade, news? Well, because the only difference between a warm, oxygen-rich corpse and a warm, oxygen-rich *you* is the ongoing cellular processes repairing your body—Builder demons, if you like. If I stop a cellular repair process with cyanide, or slug bait, everything decays very rapidly. That means, of course, the rate of cellular repair during normal life must be equally rapid. If you are always decaying rapidly, you must be also getting repaired rapidly.

---

*This doesn't work in reality because life is very complicated, and you are filled with a swamp of bacteria that will, like an old person's cat locked in an apartment, eat you surprisingly quickly when you die.

†I mean, don't try this at home. Just satisfy yourself with the mental exercise. Last thing I need is a lawsuit where your dog/child/miniature pig ate your oyster-soup experiment.

This all stands to reason, right? Well, I hope so, because I am about to insert some extra complications.

We know how fast cells are replaced, having been studied for a long time, and most biologists think of replacement at the cellular level.[2] Some cells, like your gut epithelia or your cheek linings, replace themselves completely every four days or so. In contrast, almost none of your brain cells are replaced after a certain age. I say almost because of the long-established "fact" that none of your brain cells are replaced is incorrect. The dentate gyrus replaces cells, and so does your substantia nigra, with a very small possibility that other parts may be replaced that we don't know about. The rest of your body is somewhere in between "a couple of days" and "never"—cells mostly get replaced every month or so.

This huge diversity in cellular turnover rates, from superfast to zero, generates a problem for our little thought experiment. If repair rates must balance out degeneration/breakage rates, why does the brain rot about as rapidly as the cheek lining? That is, how is the cheek and the brain decaying at a similar rate but the cheek is replaced very fast and the brain seemingly never?

My thought experiment suggests rapid and constant brain repair to balance out the rapid decay, but clever biologists say that the rate of brain-cell replacement is zero. So, how do we reconcile this problem?

The problem lies with biologists' assumption that the cell is the only unit we need to consider. We generally think that if a cell is not replaced, the tissue and larger units are not replaced either. That's why many people talk about brain aging as if it were a static system, steadily accumulating damage over time as we age.*

Cells are made up of smaller parts. So, while my cells may not be replaced as a whole unit, we could look at the *smaller* subunits. What smaller subcellular parts is the brain, and the rest of me, made of? How often are those replaced?

If I think too small, I find atoms are tiny, easily replaced, and functionally indestructible—unless I nibble some plutonium or step into a particle

*Many biologists have rightly wondered, How do brain cells endure for eighty years without being replaced? Surely, if these cells are there from maturity to death, they can accumulate damage like a worn-out machine that *also* never has its parts replaced?

accelerator at the wrong time. But what about in-between sized parts, bigger than an atom but smaller than a cell?

At this size, organisms are made of molecules: proteins, lipids (fats), a very low number of minerals and salts, a small amount of DNA, and RNA (a half-strand of DNA) but mostly water—which is very robust and easily replaceable like atoms. While all bodies are different, of course, let's get a general idea of the numbers of each type of molecule in a cell. By weight, a typical human cell is approximately:

Water—65 percent

●●●

Proteins—20 percent

●●●●●●●●●●●●●●●●●●●●

Lipids (fats)—12 percent

●●●●●●●●●●●●

Other inorganic molecules—1.5 percent

●•

RNA—1 percent

●

Other organic molecules—0.4 percent

•

DNA—0.1 percent

·

The percentage of inorganic compounds will be higher in a whole person—mostly because bone minerals like hydroxyapatite (a crystal) and some salts do not sit inside your cells—and everyone is a little flexible on the fat content, but you get the idea.

The next question is, then, how fast do each of these subcellular body parts break down?

Water is uninteresting. Like the inorganic salts further down the list, these molecules are replaced with perfectly identical molecules because you drink, sweat, and pee, not because they're broken and need repair. Boring.

In contrast, proteins are wild. Due to very recent advances in protein chemistry, we now have a very good idea of how long proteins last. Most proteins in your body have a half-life measured in minutes.[3] A few more have a half-life of hours. A vanishingly small number have a half-life longer than a few days.[4] Collagen is perhaps the standout, with a half-life in the body of about 15 years for skin collagen or 117 years for other collagens.[5]

Despite the long half-life, even collagen replacement rates are fast enough to be seen on short timescales. Scars seem like fixed features of your body, yet they are constantly being repaired and replaced while maintaining the same shape, using collagen fibers that have a slightly faster rate of decay than typical skin collagen.[6] We know this for a rather grim historical reason: people in sailing ships kept diaries and got scurvy. In the terminal stages of scurvy, your body loses the ability to make new collagen and therefore cannot maintain or repair scar tissue. Old wounds will literally reopen and unzip themselves along the scar.[7] Even with the long shelf life and highly robust nature of collagen molecules, tissues made of collagen (scar tissue) dissolve fast enough that should you set sail from Ushuaia without a barrel of lime juice, your collagen will decay before you reach Tahiti.

Most of your proteins—which constitute one-fifth of your body—will be replaced by the end of the week. Even your scars and wrinkles get replaced during a slow sailing trip. Protein-wise, you are a little baby, freshly made by the organic recycling machinery in your body.

Okay, so what about the lipids (fats) that make up 12 percent of my body or, when I get too close to ice cream, a few percent more? Lipids carry out hugely important tasks in your body, perhaps the most important of which is forming biological membranes, including cell membranes that keep your cells intact. We can now measure half-lives of cell membranes and their component lipids. Like proteins, lipids seem to break down extremely fast, but these rates are very complex to estimate.

Because it's hard to get ethics approval to put radioactive tracers into human brains, even harder to get volunteers, and hardest of all to mash up the volunteer's brains to extract a sample, most experiments on lipid turnover are done in rats and mice. Here, while "the half-life of phospholipids [fatty molecules that make up the cell wall] is consequence of a constitutive recycling of membrane phospholipids by a complex biochemical system," it seems that the "half-lives of fatty acids in individual phospholipids are minutes to hours" in a living brain.[8]

The energy output of your brain reflects this reality. In the brain, about 1.4 percent of a typical mammal's daily energy budget is used for brand-new phospholipids, and a further "5% for recycling of fatty acids within phospholipids, 7.7% for maintaining membrane asymmetries of charged aminophospholipids, and about 12% for maintaining [or making new lipids used in signaling]."[9] Given that the brain takes about 20 percent of your energy budget, that means about 5 percent of your daily energy is spent replacing lipid [fat] molecules in your brain. One out of every twenty calories you eat is for your brain-fat recycling program.

The rest of your lipids are much the same. Lipids insulate the chemical wiring of your brain, which are replaced above, but they also constitute the energy-storing fat cells in your body. Fat cells are still important, though their function is largely derailed by malfunction in rich societies. These cells are wired up to be chemically stable, storing energy in a way that is robust and lower energy using a lipid called triacylglycerol. Triacylglycerol in fat cells has a much longer half-life but still only manages six to nine months in humans.[10] The white fat cells themselves, by far the most common adipose fat cell, have a half-life of maybe eight to fourteen months.[11] The least functionally constrained lipids, not used for brain insulation or signaling, are replaced the slowest at the cellular and subcellular levels.

What about inorganic molecules? Bone is the heaviest inorganic component because it contains so many hydroxyapatite crystals and would seem completely static at first. However, adult bone is replaced roughly every four to ten years: much slower than proteins but quite a lot faster than you might expect. Cells called osteoclasts (think Breaker demons) destroy bone on purpose, and cells called osteoblasts (think Builder demons) replace it with fresh bone. Importantly, this rate of breakdown and repair is not dominated by damage caused by environmental inputs like running or fistfights but by an active process regulated by your body. Your cells *decide* the rate of destruction and repair. Baseline rates of repair are very high: children's bones are replaced every year, and when you break a bone, your cells can speed up the normal repair process to fix you up in a few weeks.[12]

Curiously, the inorganic molecules in tattoos are not replaced, unless you get reinked, but they act in a slightly different and very interesting way. The ink has an extremely long half-life that is effectively negligible compared to human lifespans, so tattoo ink hardly fades. It does blur, and slowly leaks out of the tattoo and is absorbed, but that is different. So, why do tattoos last for decades if your cells or subcellular parts are being replaced so fast?

The answer is linked with how your body reacts to foreign objects like, say, the thorn I've had lodged in my cuticle bed for six months. Initially, the thorn caused an infection that tried to either push the thorn out or dissolve and eat it with special cells. When this infection response failed, my immune cells formed an "encapsulation response." Cells called macrophages created a kind of living shield around the thorn, like a prison made of interlocking glass bricks, holding it in place and sealing it off from the rest of my body.

The same process happens with tattoos. Your immune system cannot push the ink out of your skin, so macrophages form a living wall to seal off the ink and keep it isolated. This living wall of cells is replaced one by one over time, allowing some ink to leak slightly and causing the tattoo to slowly blur, but the capsule of cells stays locked in the same shape. The "permanence" of a tattoo is therefore an illusion: a permanent shape maintained by an impermanent capsule of living and replaceable cells.

After these inorganic compounds, the next most common part of your body is RNA: the single strand, message-carrying and protein-making threads of half-DNA that are produced in extraordinary amounts by transcribing or "reading" your DNA as a reusable template. The half-life of RNA is complex and highly variable, partly depends on the cell cycle, and partly depends on uncharacterized differences across thousands of types of cells and cell conditions in your body. However, a good ballpark estimate for how long RNA lasts in human cells in a lab is half-life of about ten hours.[13] About 5 percent of RNA has a half-life of less than two minutes.[14] Again, it seems possible to have very fast construction and replacement of subcellular body parts. Damage to RNA also seems largely unimportant because it decays so fast and is replaced at a furious rate from a (more or less) fixed DNA template. It's as if you were printing off new versions of RNA from a saved electronic file at the maximum speed of an industrial printing press. Errors from typos in the electronic file are more important than the paper breaking down.

We have worked our way down to the second to last category of "other organic molecules" at 0.4 percent of your body. These are such a mixed bunch that I don't know where to begin, but they generally seem to follow the same dull category as water: they are ingested and excreted rather than repaired.

Finally, we have DNA, which has an extremely long half-life and an extraordinarily complex machinery that repairs it at all times of the day. As a refresher, DNA is a (very strong) double helix molecule that resembles a ladder twisted into a spiral, where each rung in the ladder is made of two halves: a pair of letters stuck together like two puzzle pieces. Each letter only attaches to the correct matching letter on the other side—a G matches with a C, and a T clicks together with an A.

Scientists have estimated the half-life of the individual letters that make up the twisted-ladder double helix. Some numbers are truly incredible. At a body temperature and pH of 7.4, a bloke named Lindahl measured "a half-life of about 30,000 years for each cytosine residue" and about 400 years for cytosine in RNA.[15] Even outside the body, where no repair mechanisms exist and the stabilizing structure of the chromosome breaks down, DNA has a half-life measured* in centuries.[16] This is why we can still reassemble Neanderthal, mammoth, or saber-toothed cat DNA from partially decayed fragments—a discovery that won a fantastically well-deserved Nobel Prize.

In addition to DNA being a bulletproof information-carrying device, it is also actively repaired in the body using an army of molecular machinery. Each of your cells contains about 3.2 billion letters in the genome, arranged into two matching strands. To extend our ladder-puzzle metaphor, each rung of the DNA ladder has two halves, which only click together like a puzzle piece when matched with the correct half. This means that if one letter is destroyed, only the correct letter can fit back into the gap that is left, allowing an exact replacement.

Around 2,000–10,000 A/G letters in your genome are repaired this way, and about 20–100 times fewer Ts and Cs, in each day and in each cell.[17] These single-letter repairs are so accurate that you only accumulate 20–30 or so new purine mutations (incorrectly repaired damage) each generation, including mutations caused by other damage processes. Because there are

*We know that DNA may not strictly have a "half-life" outside the body because death causes purines to fall apart rapidly at first and more slowly later, longer strands of DNA break at different rates than shorter strands, and DNA decay rates may not follow something called first-order kinetics.

roughly 8,000–10,000 days per human generation, this means there is at most one failure for every 2 million or so single-letter repairs.

Builder the repair demon is doing a spectacular job. But this single-letter "base excision" repair is just one of your many DNA repair mechanisms. When longer strips of letters are destroyed on one-half of the DNA molecule, repair is accomplished using another repair process, using a "puzzle-piece" matching process called long-patch base excision repair. This process is also extremely accurate.

More dramatic double-strand breaks of DNA happen when the backbone "ladder" of DNA is broken completely. Both halves of the DNA ladder are shattered, not just one rung, and there are no complementary puzzle pieces to fit back together. Because DNA is so robust, this kind of damage is much less common and requires a direct hit on DNA from ionizing radiation or something similarly drastic. While information is lost with double-strand breaks, the DNA is repaired and the gene usually continues to function normally.

I could continue, but rather than list all these repair mechanisms and their accuracy rates, there is a simpler way to observe the half-life and damage processes. An amazing part of the entire DNA story is that we know the total "lifetime damage" that is irreversibly included into the human genome, from all causes of damage put together. We could use theoretical estimates from the utterly brilliant scientist J. B. S. Haldane, who we will meet in the next chapter, or we can measure the rate directly.[18]

Of the 3.2 billion letters in a human genome, about 40 are damaged and repaired with the wrong letter per generation. Forty, from your generation as an egg cell to the birth of your own children decades later. There are other forms of irreparable damage in your DNA, usually with odd names like indels and stuttering-end repeats and chromosomal inversions, but these are even less likely. Amazingly little of your DNA is ever damaged, and there is no indication that this highly random damage causes aging.

An important repairable system clings to the outside of your DNA. All our cells carry the same DNA yet do different tasks: one cell is a liver cell, while another forms Meissner's corpuscles. One way cells achieve this remarkable diversity is to have inherited changes that alter how DNA operates under different conditions. These are called "epigenetic" modifications. Epigenetic signals are a bit of an unknown. They can persist across generations and therefore have the capacity to inherit early life damage and carry it forward. There are even "methylation clocks" that can estimate human ages based on epigenetic profiles: we talked about these methods much earlier in the book, and they are a pretty exciting prospect to improve aging research.[19] Therefore, the risk is that these little marks accumulate damage over time. They *may* accumulate damage, but we don't know. We are still discovering basic things about epigenetics, and our knowledge is very limited. Some epigenetic "switches," like the histone methylation marks, have a half-life of about six months to a year; others can be deliberately erased and overwritten, but most are hard to measure. Others still are a mystery and were discovered only a few years ago.[20] The story here is that we have an unknown, though some systems (like histone methylation) must have some means of repairing themselves if they have a six-month half-life but persist for generations.

Finally, we reach the only parts of your body that are not replaced. It's a short list. Teeth, which are the evolutionary descendants* of boneless fish scales, are not replaced in adults.[21] They could be replaced if evolution wasn't a harsh mistress: apes could easily have evolved six sets of adult teeth like elephants, or endless replacement teeth like rodents, and be free of our dentists forever.

The only other part of you that isn't replaced is two layers of your corneas, the light-refracting layers over the lenses in your eye, that are around a fifth as thick as a human hair. Unlike the rest of your eye, only the corneal endothelium and Bowman's layer seem to never be replaced during your life. Of course, this again seems to largely be a whim of evolution: these tiny layers *could be* replaced, but they seem to last long enough that it doesn't matter. Even weirder, Bowman's layer may not need replacing because it doesn't seem to do anything and appears to be a biological residue left over from early development.[22]

In short, the body has an extraordinary capacity to repair and replace itself on a small scale. A few key elements of your body may be essentially indestructible, like water, or can break down but are constantly flooding in

*Shark skin preserves the original evolutionary function of our teeth and, thanks to their shape and form, are known as denticles, meaning "little teeth." In sharks, teeth are just specialized denticle scales that grow inside the mouth. In humans, amazingly, they seem to be the same thing: specially shaped fish scales that we use for eating.

and out of you anyway, like CO_2. Other components of your body, like the components of your teeth and DNA, are near-indestructible but could be repaired or replaced if needed. The rest of you is decaying and being replaced at a *shockingly* fast pace. The only primary* kind of damage "carried forward" occurs in your DNA, and the vast, vast majority of that is repaired. Even the permanence of your wrinkles and scars is an illusion: they are being replaced, not accumulating damage, and they hold their shape because your body *makes* it that way. For proteins alone, over one-twentieth of your weight will be dead, recycled, and replaced by tomorrow.

To put this in terms of our Buddhist story at the start of the previous chapter, it is as if Breaker will rip off your arm, and Builder will dissolve the dead arm, reconstitute it to a new arm, and *stick it back on*. This happens—at a molecular scale smaller than your cells—every day.

The complex process of cell recycling may be slightly faster or slower for different components, but even when the rate is slow, this tells us something more profound. Your body can effectively replace broken parts, even if they only last a minute or two, whenever it needs to do so. The reason collagen or teeth are replaced slowly is not that you cannot replace or repair these parts indefinitely. Instead, your body does not need to replace these hard-wearing parts: it could evolve to do so, but it isn't necessary.

The same logic plays out at larger scales. Cell replacement rates reveal that cell and tissue replacement can happen in days—as it does in your gut lining or cheek cells. While it seems like the brain is a special case—neurons need to keep the same shape or your mind may slowly dissolve—everything else in your body is very replaceable at the cellular level. Even tissues that are not replaced rapidly could be replaced rapidly (in the same way that we rapidly replace the tissues of broken bones).

Many scientists learn about this concept as "somatic [meaning body] maintenance," another machine-based metaphor, and give it little thought: somatic maintenance is a cellular process running in the background, something you use to normalize gene expression or adjust your statistical values. The fact that you are being *recycled* as you snore, walk, or eat a pie is not conveyed very well.

---

*The only unknowns are epigenetic modifications, yet even with these the question has to be asked: if we can accurately repair DNA and indefinitely replace macromolecules, why not just evolve to do the same with epigenetic signals?

Very little of your body will still be the same in a month or a year: the atoms will still be inside you, but they will have been rearranged, repaired, and recycled within a cell to form new molecules over and over again.

Heraclitus was right when he said you cannot step in the same river twice because it will not be the same river and you will not be the same person.

You are a ship of Theseus, where the planks rot out every eighty-five minutes.

There are some points to the general argument that can be wrapped up, like the observation that damaged macromolecules become more common in your body with age. This has long been used by the "damage-based" theories of aging to claim that damaged macromolecules "build up" over time. Yet for proteins, consensus is now forming that there is no "accumulation" of molecular damage over time.[23] Rather, damaged proteins become more common as you get older, not because you accumulate them from earlier ages but because your body's repair mechanisms are slowly downregulated, reducing the rate at which damage is repaired. That is, the Builder demon is inhibited and gets slower over time.

The same thing seems to happen with things like lipid turnover, mitochondria, and cells. Damage to these is not increasing in old age because it "accumulated" over a lifetime. The delay between damage and repair is just getting slightly longer: two hours instead of half an hour, a month instead of three weeks. The outcome is that a higher fraction of your macromolecules (or mitochondria or cells) are damaged and awaiting repair at any given time. The damage is not accumulating from your thirties but rather from last week.

This all raises enormous problems for our biomechanical picture of aging. An obvious fatality is the various "biomolecular theories of aging" that state that damaged macromolecules (large molecules)—broken by radiation, oxygenation, or some other nasty damage process—somehow accumulate over time to cause the aging process.[24] These theories are in even worse shape than you think because while undamaged proteins have a half-life of minutes, your body targets damaged proteins* for deliberate destruction at even faster rates, using things like targeted proteases and molecular chaperones.[25]

---

*Also targeted are "extracellular aggregates"—collections of damaged molecular body junk that accumulates in odd spaces like the dust behind your fridge.

Even in old age, your body is, on average, brand-new. How, then, would a macromolecule "accumulate damage" over the eighty years of your life when your macromolecules have an average half-life of two minutes to two months? How does a tissue accumulate damage over decades when it can be replaced at whatever rate your body decides?

Take our super-slow turnover fat cells. Even starting with a typically slow fat cell half-life of twelve months and a ridiculous total of thirty trillion fat cells—roughly the number of cells in your body—we find that *every single one* of the thirty trillion cells would be replaced after forty-six years. This also means that the average age of a white fat cell in your body is twelve months: baby fat, on average. How could these cells accumulate macromolecular damage across your lifetime if, on average, they are year-old cells made of one-month-old molecules? And more vitally, if a subtype of damaged "old" molecules did cause aging, why not just replace them as we do with everything else?

Evolution has already solved the "damage molecule" problem for more than 99 percent of the molecules in your body by rapidly replacing them. Why not just do the same to any other molecule that accumulates damage? It seems a pretty small task given that the payoff is to avoid the physical costs of aging.

As we saw in the chapter on British machines, the lifespan of complex mechanical systems like cars and mopeds depends on money and choices more than "wear and tear." This raises some interesting questions given that most of the components of your body "wear out" within a day or two. But if complex machines are being repaired by the "invisible hand" of money and personal choices, what force is responsible for keeping your body repaired?

What is the equivalent to the economic payoff of repairing a car, for a human being, slug, or crocodile? What is the invisible hand in organisms that keeps them running?

This fascinating question consumed me for years.

In the next chapter, perhaps the most important, we will directly calculate the payoffs of life.

14 Darwin, Bees, and Big Family Trees

What do you gain from living?

Darwin seems like a good jumping-off point. His 1872 description of natural selection is perhaps best: "can we doubt (remembering that many more individuals are born than can possibly survive) that individuals having any advantage, however slight, over others, would have *the best chance of surviving and of procreating* their kind? On the other hand, we may feel sure that any variation in the least degree injurious would be rigidly destroyed. This preservation of favorable individual differences and variations, and the destruction of those which are injurious, I have called Natural Selection, or Survival of the Fittest" (my emphasis).[1]

Maximize your chances to survive and procreate—okay.

Darwin seemed to define the biological purpose of existence as fitness maximization—the most fit individuals are those who survive and procreate, passing on their inherited traits. Beautiful as the idea is, Darwin is almost unique among modern biologists because he used elegant inductive logic and qualitative arguments (not quantitative number- and math-filled arguments) to establish his theory. He left us essentially no mathematics to define what *fitness* is, exactly, and therefore we do not know his exact biological meaning of life.

We have a good idea, but "good" is not the same as mathematically exact.

The job of mathematically defining Darwinian fitness fell, initially, to the founder of eugenics and famous statistician R. A. Fisher.* He's an extremely

---

*Fisher is highly controversial. He incontrovertibly did immense good in the world by improving global science through his statistics: they form the foundation of medical science, much of modern statistical methods, plant breeding, and clinical trials. He also did immense harm. Fisher defended the cigarette industry from criticism for decades, including by pushing the idea that cancer caused smoking and not the

important person but not necessarily a great choice. Fisher decided, axiomatically, that "fitness" was measured by the average Malthusian parameter *m* of every individual carrying a given trait or gene. The Fisher equation for this parameter was lifted from a continuous equation by Alfred Lotka and a discrete equation from Leonhard Euler—without attribution or reference in a process known to most people as plagiarism—and the equation was defined as the measure of fitness because Fisher said so.[2] In other words, organisms existed to maximize baby output at a rate determined by the (abducted) Euler-Lotka equation.

Under this definition, the faster you made offspring and the higher your average Malthusian rate of growth, the better. This Malthusian parameter *m* is often also called the "natural rate of increase" R_0 while remaining defined by the same equations.

This was the biological aim of existence—to maximize your average *m*. In density-independent populations with discrete generation times (i.e., under simple conditions), the Malthusian growth rate is a function of the number of female children born to each woman at a given age *x* (or male children born to each man, if we follow Euler—it is supposedly an arbitrary choice which definition you use), and the fraction of the original population surviving from birth to that age *x*.[3]

Sorry for alluding to math, but this thing is important, and there won't be a test. Repair your body to avoid death, and this is the equation that determines your payoff.

In stealing/lifting/overlooking the authorship of the Euler-Lotka equation, Fisher ensured that many later biologists overlooked the assumptions of Euler and Lotka, including Lotka's idea that using males or females in this equation gives an identical result (it doesn't) and the idea that only direct reproduction of offspring counts toward fitness.*

Under Fisher's definition of Darwinian fitness, we can calculate the evolutionary fitness lost by dying at any age in any biological population.

---

reverse. He also drove forward the field of eugenics—which he shielded from criticism using a lot of bullying and (I kid you not) a slightly shit potato experiment that apparently ruled out gene-by-environment effects everywhere in all environments and all species.

*This idea carried directly into Nazi ideology, fueling the idea that survival and procreation were all that mattered, especially of "their kind," and that nonreproductive people were genetically useless and could be disposed of. More on that in a moment.

Supposedly.* Put another way, this equation tells you the evolutionary payoff of repairing yourself and avoiding death—if R. A. Fisher is correct. Under this equation, you repair yourself to maximize the rate of *future* baby production—the biological meaning of life.

The first person to detail the full repercussions of this was a scientist named W. D. Hamilton, in a remarkable mathematical proof from 1966.[4] Hamilton showed that if fitness was measured by the Euler-Lotka equation, the strength of natural selection must either remain constant or become smaller with age under all conditions. Even if you began with a potentially immortal, ever-breeding, infinite-food-having population, the selective costs of any death would become smaller with age.[5] The upshot? That the evolutionary cost of dying is at best constant or, once you start having babies, gets smaller with age *in every possible living system.*

Under normal circumstances, the evolutionary cost of death reaches zero as you reach the last age at which anyone in a species reproduces—beyond this point, there is no benefit to living and no cost of dying. As the evolutionary cost approaches zero, random noise (in a process called genetic drift) becomes more important than evolution and takes over. Unlike the exponentially growing payoffs of, say, repairing a classic Ferrari, this means that the future evolutionary payoffs of repairing your body *always* diminish with age and approaches zero at advanced ages.

Hence, aging.

Hamilton's work is a very singular paper—a mathematical proof that aging must always happen, in a universe where fitness is measured by a combination of offspring production and survival, is a hell of a thing.

This particular framing of "fitness" is known as direct fitness theory. However, there are a few problems.

Surprisingly, even as Hamilton laid his ideas down on paper, he recognized that his proof of aging was missing something vitally important about evolution.[6] Thanks in part to work by earlier biologists, Hamilton knew that menopause and postreproductive survival were serious problems that were not, and could not, be predicted when fitness was strictly defined by the "optimize making babies" definition of fitness.[7]

---

*As an undergrad, I once gave a simple demonstration of why this was not true in certain species of ants, which had either zero or undefined fitness under the Euler-Lotka equation, and was firmly told to shut the f—up.

Haldane's proof details that the evolutionary costs of death should be constant at any age where an organism is not reproducing. This includes the idea that before puberty, the fitness lost by dying should be constant (you lose all future reproduction), and after menopause the evolutionary cost of dying should be zero. Under Fisher's direct fitness theory, it was therefore impossible that an organism would evolve a mechanism to die or become infertile on purpose, foregoing all future reproduction. An organism should never, and indeed *could never*, evolve to deliberately stop reproducing: you cannot maximize the product of fertility and survival and give yourself the "best chance of surviving and of procreating" by deliberately making your future fertility zero.[8]

But what, then, is menopause?

A carefully regulated and evolved mechanism for halting reproduction in humans, universally conserved and expressed by every woman on earth who reaches a sufficient age.

The extensive postmenopausal survival of women is hugely problematic for direct fitness models. Even in high-mortality hunter-gatherer populations, a substantial number of women (~40 percent) survive to reach menopause: it is not a modern statistical anomaly.[9] Even harder to explain is the fact that the rate of reproductive decline happens early and continues more rapidly than other systems because of evolved and tightly regulated* physiological differences.[10] If evolution is supposed to maximize survival and reproduction as a single unit, it is not clear why reproductive capacity declines much more rapidly than the biological systems that keep you alive. However, this rapid systemic decline of female fertility is universal and appears tightly regulated across all humans, suggesting† a strong role for adaptive evolution.[11]

The evolution of deliberately *not* reproducing should be fundamentally impossible if fitness attempts to maximize the Euler-Lotka equation. This

---

*In evolution, if something is "functionally invariant" (always happens in the same way and takes the same form), it is usually extremely important; otherwise, random chance would cause the trait to "drift" randomly over time. Menopause seems conserved to such an extraordinary degree that there is no known case of a woman simply *not* undergoing menopause and reproducing indeterminately. That is remarkable and suggests a strong and functional evolutionary role for it.

†Though there are some people who argue that menopause is a nonadaptive (incidental) process.

should make selection for menopause, and any postreproductive behaviors, or anything that kills you or makes you sterile. Yet menopause seems to have independently evolved in species as different as toothed whales, carpenter bees, and social aphids—whose grandmas become "wax-packed chemical warriors" defending their family from threats as "living bombs of glue."[12] Quite a retirement. Given the difficulty of studying entire reproductive lifespans in the wild, I suspect we are yet to discover many other species that have evolved menopause or a similar strategy—for example, it took us six decades of field studies before we first detected signs of menopause in wild chimps.[13]

Hamilton saw postreproductive survival as equally problematic to a question that has received much less attention: prereproductive survival patterns. Hamilton wondered, If the Darwinian payoff for survival is at a constant maximum value before you start to reproduce, why are survival rates so variable across childhood and infancy? That is, why are heritable causes of death so unevenly distributed—with babies dying from heritable Mendelian diseases at dozens or hundreds of times the rate of two-year-old children or second trimester fetuses? Even in the protected environment of the womb, mortality risks are hugely uneven over time. Evolved, heritable, Mendelian diseases have always been highly uneven throughout early life and are fast becoming the number one cause of death in infants across the developed world. So, the question of "why would evolution allow such an uneven mortality risk?" is hardly academic: our ongoing battles in neonatal medicine depend on understanding such genetic diseases. But we still don't really understand this one at all.

Initially, I found this the weaker of Hamilton's two arguments. Perhaps such differences were, I reasoned, caused by the hugely different environments of the womb, birth, and infancy. Perhaps they were caused by genes being turned on for the first time during development, and failing or causing disease, and creating nonlinear mortality risks over time. Perhaps, but this ignores something else. In other species, selection and evolution has shaped the risk of death so that infants are the least likely to die in the womb, nest, or egg. So why is this not the case in humans?

We don't know.

These are just the objections to the Fisher/Euler-Lotka equations in human biology. Yet evolution is not a special case: it doesn't make exceptions for humans. We are only one species, and even so, our case throws up major problems for the idea that organisms evolve just to maximize a "survive and reproduce" equation.

If there is one takeaway from studying biology, it is that you can't generalize from humans to other species—we are either too weird or, usually, not nearly weird enough. The problems observed for Fisher's "fitness is maximized reproductive output" are far more substantial in other species,* especially if we consider the other thirty million species that routinely evolve very different patterns of mortality.

Darwinian fitness is supposed to be universal across the tree of life. Is it?

In Fisher's terms, no.

Infant mortality is one example. While mammals generally have risky births, for other species, birth is the lowest-risk part of their prereproductive lives: birds, for example, usually suffer the highest mortality risk during young juvenile stages when they fledge and leave the nest. Ants suffer extremely low mortality rates as infants in the protected environment of the colony but die at atrocious rates once they start work as foragers, hunters, warriors, farmers, mortuary worker specialists, construction engineers, wasp guards, slave raiders, living honey jars, or any of the literally hundreds of jobs they take up on maturity.[14] Evolution *could* evolve lower early life mortality risk in mammals, as it did for infancy and early childhood, but it has not done so.

Darwin recognized another early problem in worker bees, which (he thought) do not reproduce. If menopause is an evolutionary riddle, sterile worker bees are a disastrous conundrum. Darwin wondered, Why do worker bees evolve to *avoid* their own reproduction and cooperate? Why do they expend their lives for the reproduction of others?

The mystery is even deeper than Darwin may have appreciated. Thanks to something called "haplodiploidy," honeybee workers are capable of immaculate conception, where they develop ovaries and give birth to males without having sex. Usually, they do this after the queen dies, and there is a brief period of anarchy where every worker uses the colony's leftover resources to have as many children as possible. However, there are also—and

* We have only recently discovered the astronomical number of genders in fungi, for example, which really messes up the assumptions of the Euler-Lotka equation that "both" genders are interchangeable in the equation. There are dozens to thousands of mating types/genders in common fungi—and we have only looked at some.

I love this—"anarchy phenotype" honeybees, which secretly have babies despite nominally serving the queen.[15] In other words, worker bees have all the equipment, genes, and biological tools they need to reproduce. Yet they don't. As a result, the riddle of why honeybee workers give up reproduction for the queen is a classic problem in biology. One that, no matter how much you twist it, does not fit into the framework of treating individuals as personal baby maximizers. You should not be able to evolve to *not have babies* if baby maximization is the meaning of Darwinian existence.

This problem keeps getting bigger the more we explore. Thousands of species—including ants (>12,000 species), bees (>380 "advanced" eusocial species), wasps (~1,100 highly social species, a few with sterile workers), a tiny insect called thrips (a few species), an odd collection of shrimps, some trematodes, one beetle, and termites (>3,000 species)—have all repeatedly evolved sterile workers or warriors through independent evolutionary events. For perspective, there are only around 6,400 species of mammals—mostly Rodentia and Chiroptera, or rats and bats. Yet even the mammals are not safe from exceptions. Two species—Damaraland mole rats and naked mole rats* (which look like a weasel with male pattern baldness)—were only recently discovered to have evolved voluntarily sterile workers.[16]

Evolving sterility should be a contradiction. Yet it happens, over and over, and seems strikingly successful—the total biomass of ants alone, most of which are sterile, "exceeds the combined biomass of wild birds and mammals."[17]

While Darwin and Fisher (and many others) propose an explicit purpose of life as direct reproductive output achieved by maximizing survival and reproduction, we know that many organisms have evolved ways to stop reproduction or halt survival.[18]

Halt survival—does anyone evolve to do that?

Inside multicellular organisms, there is a way to kill off dysregulated, unruly, toxic, or cancerous cells called "apoptosis." Everyone learns about

* Mole rats are studied for aging because they have very long lives, but this is perhaps the least interesting aspect of their lives. Mole rats have a colony song, similarly to how ants have a "colony smell" inherited from the queen, and they learn and "chirp" a password to enter their home.

apoptosis in undergraduate biology, and for a multicellular organism the idea makes sense, but one cell is not the death of an entire organism. However, one researcher named Vladimir Skulachev claims that apoptosis also happens in (and evolved in) single-celled organisms like yeast.[19] Yet we don't need to quite look as far as Skulachev's work—which seems, let's say, curiously unreplicated. We just need to remember basic bee stuff, like that honeybees die when they sting you. In evolutionary terms the kamikaze death of honeybees is completely avoidable. The basal or original state of bees is an ability to sting repeatedly without dying, which many bees and their wasp ancestor cousins do with happy gusto. Honeybees evolved their sting-once-and-die approach even though it means they also evolved to lose their chance for any post-sting reproduction.

Honeybees are not the only programmed, evolved suicide-like behavior. Ants somehow generate volunteers that stay outside the nest before thunderstorms and seal up the entrances—with 100 percent mortality rates.[20] Even the menopausal glue-bomb warrior aphids have evolved to die for a cause.[21] Ants, bees, and wasps also routinely do jobs that, while not exactly suicidal, have extremely short survival times (like mortuary workers).[22] Yet perhaps the coolest suicidees are slime molds like *Dictyostelium* that, delightfully, go from a free-swimming single cell to aggregating and merging their cells together like a gigantic sci-fi swarm, before 20 percent of cells in the blob kill themselves to make a death-pile.[23] The death-pile marginally helps the rest of the blob disperse to new lands.[24]

All of which raises a question: if you are striving for the survival of the fittest, how do you evolve voluntary death?

Even more mathematical issues spring up in different species if you look for slightly different problems. I made some short, unpublished mathematical forays into the field showing that death is often selectively neutral under the "maximize survival and reproduction" equations. That is, you don't evolve to die, but survival is unimportant for fitness. This happens when survival and reproduction become statistically independent from one another. For example, the thirty thousand or so species of snails, including common garden snails (*Cornu aspersum*), are hermaphrodites and fire "love darts" (horrifically large harpoons) into one another before ejaculating. This trick allows for the incredible long-term survival of sperm, independent of the survival of the snail. Snails therefore routinely reproduce long after they die. A similar thing happens in (male) eusocial insects, all of which die before reproduction in many species. This disrupts the equations above because reproduction is not tied to survival: the "survival of the fittest" equations don't work if you are dead and reproducing.

Like the answer for evolved sterility, the answer to evolved death pathways requires that we either do a huge amount of mathematical fudging or throw out Fisher's definition and try again. So how do we solve this problem, in a unified way, across the entire tree of life?

Unfortunately, I don't have a good answer, only my own approximate guess. I spent five years of my life figuring it out with the help of my brilliant PhD supervisor (but I'm not a genius) and trying to find other people who had found their own solution.[25]

Let's start with Darwin again. Darwin proposed a solution to his own worker-bee riddle, which he saw as "the one special difficulty, which *at first* [my emphasis] appeared to me insuperable and actually fatal to my whole theory."[26] He gave an alternative theory that, essentially, sterility could evolve if sterile workers helped their reproductively active relatives to have babies. It falls short on the details, with no fault to Darwin, but in general it is coherent and an excellent foreshadowing of the way things would go.

Then we can add some more popular foreshadowing from a mustache-twirling genius named J. B. S. Haldane, a man who thought WWI was fun and who came home from the front line so that his equally mad and brilliant father—who literally invented putting canaries in coal mines—could lock him in a homemade gas chamber (filled with chlorine gas) so his father could (re)invent the gas mask. Haldane felt that being gassed was rest and recuperation and went back to the trenches after he got bored. Haldane is immensely fun as a person: his exploits included riding a bicycle across no-man's land on the Western Front in WWI to wave at the German machine gunners, carefully explaining why Hitler and the entire German nation were full of disease-carrying mutations in the 1930s (and were therefore dunces who knew no genetics) with an implication they should start their eugenics program by shooting Hitler, warning everyone to stop dunking on refugees and start protecting Jewish people in 1931, renouncing his British citizenship to become an Indian national in the 1950s as a protest against imperialism, teaching biology at Cambridge despite never having any biology qualifications, and making an endless stream of scientific discoveries.

Haldane was one of the three architects of modern genetics. Yet when he wrote a book chapter titled "What Is Fitness?" two years after Fisher had defined fitness, Haldane clearly didn't agree with Fisher: he cited Fisher's book in the chapter but did not provide any definition of fitness.[27]

Haldane instead discussed ideas that would later be requoted by biologists almost ad nauseum. His quotable quip was, apparently, that he would only dive in and "lay down my life to save two brothers or eight cousins" from drowning.[28] As Haldane shares one-half of his genes with two brothers and one-eighth with eight cousins, he would only risk one Haldane to save one Haldane's worth of genes. Due to the obscure venue where Haldane was quoted, most academics copy-paste the quote and never see the rest of what he said: that "on the two occasions that I have pulled possibly drowning people out of the water (at an infinitesimal risk to myself) I had no time to make such calculations."[29]* Haldane then gave a more complete description that calculates, quite exactly, the conditions under which a gene can become more common, even if it raised your risk of drowning "by one in ten" because it made you dive in rivers to save a drowning child who shares your genes.[30]

Haldane understood that fitness is not necessarily survival maximizing, and that it can depend on explicit calculations of passing on the most *genetic* material instead.

At the time, he calculated that the spread of such an "altruism gene" depended on multiplying the risk of an action by genetic relatedness: diving in the river would always risk your death, but it would also spread your genes if the child saved was your biological relative.[31] As a result, any gene causing an action that passed on more genes would be selected and spread throughout the population, even if it kills you.

A possible solution to Darwin's bee riddle was therefore to stop treating organisms as a baby-maximizing machine and instead treat them as a gene-maximizing machine.

---

*That is, Haldane recognized a fundamental problem with inclusive fitness, in the same way that Darwin recognized a problem with Darwinian fitness: altruism often happens without necessarily caring, knowing, or calculating who you are related to, or by how much you are related. Game theory has a lot to say about this, most of which can be applied to either definition of fitness. Having also saved two people from drowning, I can confirm that any considerations of game theory, kinship coefficients, or genetics are ignored when fighting panic and fear.

While Haldane was thinking this out loud, two different measures of fitness were being brought together and synthesized by Fisher and another part-rival professor named Sewall Wright: the other two architects of modern genetics. The first definition of fitness was obviously the Euler-Lotka equation, but the second was the definition of genetic fitness.

Because it did not have to dovetail with Darwin's qualitative logic, instead having to agree with Reginald Punnett and Gregor Mendel, the genetic definition of fitness was mathematically rigorous from the start. The thing was, fitness was only ever defined for a gene or, later, for a trait—not an individual. This was a hard task. Even defining or modeling the fitness of a single gene, in heavily idealized theoretical populations, had plagued the world's most capable biologists and mathematicians for decades and involved hugely counterintuitive results like the Hardy-Weinberg equilibrium and (later) genetic drift.[32] Even now, very few people even attempt to reconcile gene-centric and wider-scale definitions of fitness. It was therefore a considerable and seemingly unachievable task for the early 1900s geneticists to scale up the mathematics of single-gene measures of fitness to the level of individuals, which contained a then-unknown multitude of genes.

Rigorously mathematically defined by other prominent scientists, in fits and starts, the definition of genetic fitness had been shaped around the rate of change in gene *frequency*. Essentially, a gene (or trait) was more fit if it became more common in a population over time and less fit if it became less common over time. There are various ways to measure this rate of change, including the completely awesome-sounding measure of fitness called the Lyapunov exponent, but the core idea is that the faster the rate of increase, the more fit the gene (or trait), and vice versa.

In the 1930s a compromise was reached between the reproductive and genetic definitions of fitness when Sewall Wright stated that the two could be unified between population fitness (defined by Fisher) and genetic fitness.[33] Fisher, Haldane, and Wright reached a kind of accord in this "modern synthesis." A key part of the synthesis was the assertion that genetic models of fitness would be equivalent to Fisher's definition of fitness at the scale of individuals or populations, and therefore that fitness measured by the Euler-Lotka equation and genetic fitness were one and the same.[34]

So, what does genetic fitness have to say about the payoffs of living (or, if you prefer, the strength of selection for increased survival)? At first glance, genetic theory said much the same thing as Fisher and Hamilton.

The genetic force of selection, against a "fatal gene," declines with age.[35] In some circumstances, largely hypothetical ones, there may also be a trade-off between genes that raises your early reproduction and lowers your later survival.[36] However, that is about where the similarity ends, and these arguments seem to mix definitions of fitness between the individual (survival) and genetic. Indeed, the fitness terms used in these equations really seem to mix up definitions by measuring the payoffs of a genotype in terms of reproductive output.[37]

As Haldane and Darwin (when he was thinking of bees) already understood, however, the genetic output of an individual is not merely a product of *personal* reproduction. You share your genes with your biological relatives—and you can affect how often the pass those genes on by, say, working in their hive or saving them from drowning. For the individual, there often was a huge mismatch between genetic fitness and reproductive fitness.

Instead of leading to a conclusion that the mathematics of Darwinian fitness needed a rewrite, however, this conflict has led to some heavy-handed mathematical fudging. New terms are added to old equations, like sticking twenty-inch rims and a turbocharger on a horse and cart. The latest is something called "transfer theory," in which intergenerational "transfers" (of what?) shape Darwinian fitness to allow for postreproductive survival.[38] It's all a bit fuzzy to me, but maybe it makes more sense to other biologists. This kind of approach is typical—as Kuhn pointed out, elaborating on what you have is easier than revolutionizing and starting again. So, the mathematics of Euler and Lotka has gained ornaments and spawned lookalikes such that we have a few different definitions of direct fitness that say or measure roughly the same thing.[39]

There is only one core exception, which we are still attempting to formalize and understand.[40] This is the theory that Haldane seems to have worked out but never published: inclusive fitness.

In 1964, two years before he published his "proof of aging" paper, W. D. Hamilton stated that Fisher's mathematical models of evolution "admit no possibility of the evolution of any characters [i.e., characteristics or traits] which are on average to the disadvantage of the individuals possessing them. If natural selection followed the classical models exclusively, species would not show any behavior more positively social than the coming together of the sexes and parental care."[41] What Hamilton meant by "disadvantage" was the Darwinian metric of, effectively, leaving less descendants.

Following this logic, Hamilton then went on to do what so many have either lacked the capacity to do or have not been foolish enough to try.[42] He created a completely new measure of *non-Darwinian* fitness, which he called inclusive fitness.[43]

Inclusive fitness is an explicitly genetic affair in which individuals would try and maximize genetic output by any means available, including by modulating the genetic output of their biological relatives. To note, this is maximization at the level of the individual, not the population.[44] At the population level, it is arguable if fitness is maximized at all, which is a bit semantically confusing.[45] This process, where you affect the reproduction of biological relatives, is called "kin selection." This difference between Darwinian fitness and inclusive fitness is profound and has still not fully worked its way into the mathematics of evolution or population genetics. Indeed, we are still working on the first step, a formal mathematical logic of Darwinian fitness[46]—and by "we," I largely mean the singular Professor Alan Grafen—before trying to resolve how Darwinian theory integrates with kin selection and inclusive fitness.[47]

Yet Hamilton's idea already provides new ways to solve problems. One first step was to provide a more formal answer to Darwin's bee problem: individual worker bees can pass on more genes by helping the queen, with whom they share a large fraction of their genes, than they can by reproducing by themselves. This maximizes their inclusive fitness while losing all the direct (reproductive) fitness they would have gained by making their own babies. Like Haldane's "diving in a river" gene, the worker bees can spread their genes by helping relatives.

Other problems are resolved by kin selection too. For example, some of the first people to really move beyond theorizing and begin to test evidence for inclusive fitness were a mix of anthropologists and medical scientists who tested for evidence of kin selection in humans.[48] Their discovery was that postreproductive women seem to have huge positive benefits for the survival of their grandchildren, supporting the idea that kin selection is observable and plays a substantial role in human evolution.[49] This finding validated Hamilton's prediction that postmenopausal survival in women had positive survival effects on their descendants and is widely understood as a deeply important finding.[50] Treating organisms as gene maximizers, at the potential cost of their own reproduction, makes an accurate prediction that direct fitness cannot.*

Other predictions have been tested, and while there is a lot of static from proponents of game theory, the idea of organisms as individual gene output maximizers has caught on and remains extremely interesting.[51]

Inclusive fitness and kin selection are at the forefront of our slow progress in evolutionary biology and raises the exciting possibility that human survival is shaped by other forces than personal reproductive output.[52] In fact, inclusive fitness gives us a surprising pathway to explain many behaviors that seem intractable under Fisher's 1930s eugenics-style definitions of fitness where everyone is a baby factory.

If an organism can succeed by helping family members reproduce, without reproducing themselves, a considerable number of behaviors become compatible with evolutionary thinking.† Homosexuality, altruism, unconditional reciprocity, suicidally stinging that bear to death for the hive: all these things can be beneficial under different models of inclusive fitness. Almost anything is possible at this stage of our understanding. It is an exciting thought.

While filled with promise, we have barely begun to explore the potential of kin selection. Part of the problem is that we do not have the equations, and the other is because we really have problems measuring a biological family.

To really understand how diverse a set of behaviors is allowed by kin selection, let's have a look at a basic question. How many (genetic) family members does an average organism have? Thanks to some amazing theoretical results, we know that biological relatedness to family members will be higher than the genetic relatedness of any random pair of individuals, out to about 16–20 degrees of separation in your family tree. With modern genetic testing, we can even detect biological relatives out to about 12 degrees of relatedness.[53] But how many relatives sit within 12–20 degrees of biological separation from you in your family tree?

*Unless you do a huge amount of that mathematical fudging I mentioned.

†Not that they need to be. There is no law that everything we must do has to have some evolutionary purpose or grounding. Humans can exist without justification, and there does not need to be any evolutionary explanation for (say) picking your nose or painting a watercolor. Life can exist irrationally, and f—anyone who says otherwise.

This question raises one of my favorite counterintuitive results for biology. Because even for low- and slow-breeding humans, the answer is unexpectedly *massive*.

Calculating the number of direct ancestors is easy in sexual species with two sexes.* For every degree of relatedness you go back, the number of ancestors doubles in an outbred (non-inbreeding) population. Easy enough. Two parents, four grandparents, eight grandparents, and so on. It is also easy enough when you go down the family tree: the number of children per parent averages two, over the long term, so that it also doubles with every degree of relatedness. Average two kids, four grandkids, and so on.

This exponential growth in relatives, into the future and past, barely makes much of a difference for us. We don't survive long enough for very many generations to overlap. However, other species do: slow-breeding humpback whales[54] have children at age five and live to well over fifty, overlapping with eleven generations or more, and so they can interact with a truly massive number of ancestors and offspring.† For humans, however, grandmothers interacting with grandchildren is about as far apart as it gets.

Going up or down the family tree is predictable and intuitive. Going sideways is where things get weird. For example, you may think that if the average woman has two kids, then the average number of siblings must be one. Since every mother has an average of two children, there would have to be an average of two children per family so that everyone has on average one sibling. Right?

You'd be dead wrong, because the average number of offspring per woman (or per man, if you like) includes every woman who had *zero* kids.

Can you be a child to a mother who has zero kids? No.

Can you be a child in a family with zero kids? No.

As a result, the average number of children per family is virtually always larger than—and very rarely equal to—the number of kids born per person.

This has some extremely weird effects. Imagine, for simplicity, that we have an even number of women, say one hundred, and exactly half of the women have zero kids. If the long-term average number of children per

---

*Again, fungi can have thousands of breeding types/sexes/genders.

†At birth they may have four to five hundred living direct ancestors, and at death they can have well over one thousand living direct descendants on average (assuming a high infant mortality rate). For a species that has a single child every two years, that's astounding.

woman* to stay at two, this means that the other half of the women who have a family must have four kids each: fifty have exactly zero children each, and the other fifty have an average of four children each.

That means that every person has three siblings, not one, on average. This assumes there are no half-siblings, of course, which would increase the number even further.

If this happens every generation, then your two parents also have an average of three siblings: six aunts and uncles, assuming no inbreeding, each of whom will have an average of two kids (either four or none) to give you an average of twelve cousins. Things are growing quick, and we are only mapping your family tree to four degrees of relatedness (not twenty).

Because the number of ancestors you have doubles with every generation you go back, so does this "average sibling number." So, in our scenario, your 16 great-great-grandparents will each have three siblings, all of whom have descendants who you are related to: follow these lines of descent back down the family tree to reach your generation, and a typical person will have 2,046 relatives on average.†

We have only reached 10 degrees of relatedness.

This incredible scaling of family sizes is independent of the average reproductive success of a population. Throughout human history, the average reproductive success of a population has never gone far above (or below) two children per adult. Yet even in a low-fertility species like humans, the average size of our extended families is *huge*.

Making this effect even more striking is that the numbers above are underestimates, even in humans. The average family size in Europe around 1800, where we have records, was not four children per family but seven. That is, the average person had six full siblings (or even more half-siblings), meaning that a typical European in the 1900s would have had roughly 48 cousins, 192 second cousins, and well over 4,000 living relatives within

*For tricky mathematical reasons, in humans the average number of kids born per woman *must be* exactly two over the long run.

†For a fixed degree of relatedness, your relatives (in the same generation) will carry N times as many copies of your genetic material in your body as you do. This number, N, is set by the average number of siblings, the number of sexes, and the relatedness between siblings. So, with an average of six siblings, your relatives in your generation will carry 6*0.5*2 = 6 copies of your genome, distributed among relatives at the fourth degree of relatedness. With three siblings, the number is three copies; with six hundred, it's six hundred copies.

10 degrees of relatedness. Again, this is an average in a randomly breeding population*—the numbers can be far higher.

You can even see a demonstration of this first-hand by tracing your family tree or, for the lazy, ordering a genetic ancestry kit. When I ran a genetic ancestry kit, I was related to around 2,500 people on the platform despite a tiny fraction of the global population ever having bought this particular kit. Undoubtedly, tens of thousands more people are in my genetic family—sharing more genes with me than could be explained by random chance—who have never taken the test.

Once again, humans have nothing on other species. We are slow, low-variance breeders. In a nonhuman population, like an ant colony, it is perfectly biologically possible to have twenty million siblings on average, in a stable population where every parent only has an average of only two reproductively active children. At no point have scientists really been able to map the scale of these networks, let alone the statistical interactions within them.

We are getting close in humans: for example, I have worked on a family tree containing millions of people.[55]

The result is that the number of genes an organism passes on through children is always outnumbered, usually by a colossal amount, by the number of genes they pass on through family members.

The primary constraint on inclusive fitness is therefore not biological relatedness. Rather, it is the capacity to benefit or influence the extremely large number of biological relatives that you have.†

If a well-fed leopard catches an extra antelope, there is no practical mechanism by which it could distribute this extra food to its tens of thousands of biological relatives. If a well-fed ant comes across an extra caterpillar

---

*This never happens—thanks to something called assortative mating and to the fact that more reproductively active families seem more likely to persist long term, meaning we are more likely to descend from larger-than-average family trees. All of your direct ancestors must fall into the "had at least one child" category and have higher than the average reproductive success, which also helps stack the odds.

†This constraint is linked to something called the secondary theorem of natural selection, which means that selection for a trait, like death or relative-feeding care, is proportional to the strength with which that trait is correlated with your inclusive fitness. This means that selection can act via strong correlations on a few relatives or through weak (often undetectably weak) correlations with many relatives.

fifty times its size, it can *easily* distribute that food to thousands of its relatives by taking it back to the nest, making kin selection easier in ants than leopards. By analogy, in human society, I can distribute benefits broadly: for example, if I donate a maternity ward, I can benefit tens of thousands of current and future people, many of whom will be my (very numerous) relatives. The vital difference between the leopard and the ant is the capacity to modify life in large numbers of distant relatives. If that becomes possible, so does a rich set of counterintuitive behaviors that enrich for your genes.

This means that there is huge potential for kin selection in (unmeasured and massive) kinship networks, in ways that can reshape the cost-benefit equations of survival. We cannot observe these massive kinship networks, and yet we have only scratched the surface of behaviors and traits, like our rate of aging and repair, that may be explained by inclusive fitness. If we view organisms as gene-maximizing machines, we need to get started.

In 2015, my brilliant PhD supervisor Professor Simon Easteal and I crafted a small analysis that looked at the kinship-network-level effects of death. It was limited to a few thousand hunter-gatherers and asked a basic question: how much genetic material did individuals pass on through family members, restricted to within a couple of degrees of separation in the family tree, if they died at different ages?

The relative number of genes directly passed on by all the children who died was zero, of course, but the number of genes passed on *indirectly* through family members was very variable. Individuals who died in infancy passed on more genes through family members than if they died in childhood.[56] In fact, the shape of this "kinship cost of mortality" curve, which is an estimate of the inclusive fitness lost by dying, closely mimics the odds of dying overall.*

The same thing happened at older ages: instead of the number of genes being passed on flatlining at older ages, after a woman stopped reproducing, women who lived longer passed on more genes through their families—even though they were not having babies themselves.[57] As such, a simple metric of kin selection that treated humans as a gene-generating machine explained both of Hamilton's problems—why menopause exists and why

*I suspect, but have no way of demonstrating, that this is because of the huge amounts of food, teaching, and time invested in raising a child who dies.

infant mortality is variable and not constant. Examining close-up kinship networks in humans went some way in (possibly maybe) explaining the fitness payoffs of survival.[58]

Yet this is one small network, and the burden of proof in science is never decided by just one study. Who knows what a better analysis or a larger kinship network would find. I would be excited to find out and see if it can explain much of the rich diversity of death patterns.

To recap this slightly complex chapter, we have two answers to our initial "what do I get from survival?" question.

If I treat you as a Darwinian baby factory, the benefits of maintaining and repairing your body are determined by a weighted function of how many babies you expect to have in the future. This is not a rosy picture: it assumes things like "Sterile? Postmenopausal? Choosing not to have kids? Intent on saving the honeybee queen? Non-Darwinian." We know that this incredibly bleak picture is wrong, however, and even Darwin had a strong suspicion that this was the case.

Alternatively, if we treat your survival as being shaped by passing on genes, we have some hope. You can pass on genes through your very, very many relatives. Homosexuality, altruism, sterility by choice, menopause, and countless other behaviors can evolve as beneficial traits if we treat organisms as gene maximizers. But we are left with massive barriers to testing these ideas: we cannot even count your biological/genetic kin yet, let alone wrap up the mathematics.

Yet there is one area in which we have made, and are making, great leaps in our understanding of death: in organisms where kinship seems to play a limited role and direct reproduction is on center stage. Organisms where cooperation across generations is impossible, where nobody lives to care for the young, and where death is an extravaganza event.

15 The Salmon of Knowledge

If you are old enough, you might remember a cinematic sequence that graced your childhood television set. Crowds of salmon envelop the streams of North America. They struggle nobly upriver in their thousands, gracefully leaping in the water. As they travel long distances and high waterfalls, a casting call of predators take their toll: eagles, wolves, husky grizzly bears. The salmon heroically dodge these hazards and reach their clean gravel beds, where they mate. Finally, exhausted from their efforts, they die.

The great salmon migration of the North Pacific has long been a staple part of nature documentaries. It is a beautiful and ultimately poignant journey, taking place in one of the most spectacular settings on earth. I absolutely loved it as a kid.

Which brings us, as you might expect, to volume 143, issue 6 of *Transactions of the American Fisheries Society* (not, mind you, that cheap *Hustler*-esque tabloid called the *Canadian Journal of Fisheries and Aquatic Science*). On page 1,476, there is a riveting, thirty-one-page thriller with a very remarkable figure: the longest validated migration of an individual chinook, or king, salmon. A migrating king salmon was tagged and released in the Yukon—325 km from the river mouth at a town called Russian Mission—and caught two months later about 3,000 km upstream in the Nisutlin River.[1] It is an incredible distance in a very short time. Weirdly, however, the salmon that made this run was not much different from all the others. It was not an ultra-large super-salmon.

That raises a small conundrum. To see why, let's go to the river that original "leaping salmon" nature documentary, where the BBC filmed their seminal images of leaping salmon and hungry bears: the Naknek River drainage in Alaska. The town of Naknek sits at the mouth of this river system. It has 544 locals, a small pub, a post office, a dirt-strip airport, and a frozen gravel beach. Every year, people fly to the airport at Naknek. They meet the locals, drink in Hadfield's Bar, and go fishing in the river. The reason people fly here is that Naknek has huge salmon, which outsize and outweigh the Yukon members of the species and have greater energy reserves.

In the BBC documentary, the presenter spoke of thousand-mile journeys upstream, a marathon migration for these heavyweight superfish. Yet like many of Alaska's rivers, the length of the Nanek is relatively short. The farthest a salmon can get from the ocean here is just 170 km, just over 100 miles for the unit challenged, and the journey may be far shorter. A king salmon in Naknek may only swim 20 miles to spawn. Therefore, the problem is, Why do the bigger, fatter, faster salmon in the Naknek River die from a twentyfold shorter trip?

This anomaly gets stranger the more we investigate because if we shorten the river even further, the fish still die of "exhaustion." In streams across Alaska, salmon run tiny distances upstream before they spawn and die. On Kodiak Island, guides favor trophy salmon fishing in waterways less than 15 km long. Roslyn Creek, American River, Saltery Lake, and Chiniak Creek are filled with thousands of fish that may only run 5, 4, or 3 km upstream.

As I learned from my fellow winner of the Ig Nobel Prize, the bright and brilliant Jimmy Liao, swimming also requires so little energy input that salmon can still "swim" efficiently while dead.[2] The mystery is therefore why a salmon weighing 30 pounds dies from exhaustion after 3 miles and a one-day run, but a 20-pound salmon of the same species survives, in fresh water, to swim 3,000 miles. In the study of long-migration salmon in the Yukon, fish in the study swam up to 80 km a day.[3] The record-distance fish must have swum 50 km a day for two months. So why do larger trophy-sized fish of the same species die so rapidly, after a 3 km swim up a Kodiak creek?

It doesn't really make sense. Even I can swim 3 km without dying from "exhaustion"—though I would probably be enthusiastically eaten by bears. Explanations may be floated, like the stress induced by predators, waterfalls, or the shock of changing to fresh water from salt. Yet these proposals just raise more problems, like why salmon can endure such conditions for two months in the Yukon but only one day in a creek.

To really explore how little sense our exhausted salmon story makes, let's consider another salmon called the kokanee. Kokanee salmon, which have a wonderful scientific name of *Oncorhynchus nerka*, can perform the same fresh-salt-fresh migration as king salmon, and also die at the end. Yet they can also live entirely in fresh water, contained in one lake and never visiting the sea. They eventually swim up a freshwater creek, or a few yards to the shore, to spawn.

Kokanee are not exhausted. They haven't traveled anywhere. They are not in "shock" of any kind because they haven't adapted to salt water and then come back. They haven't even stopped feeding. Living entirely inside a lake is also, it must be said, generally a bear-, wolf-, and waterfall-free lifestyle. Yet freshwater kokanee always die after spawning, just like king salmon. So, if these salmon are not dying from exhaustion, what is killing them?

We were fibbed to as children. The salmon were not tired at all. They were killed by the only part of the process we didn't eliminate.

It's the sex. The salmon die from sex.

This is a hell of a realization for anyone researching aging, or who likes classic nature documentaries, but it has even more remarkable repercussions than just the fact itself. Salmon spawning is not just a sort of softly lying down on a bed to die; it is a physical cataclysm in which the body dissolves internally. A salmon's hormone levels skyrocket before and during spawning, transforming their bodies in a way more typical of a low-budget horror movie than biology. The salmon transfigure themselves in response to sex hormones, growing hook jaws, hunchbacks, and livid blood-red bodies.

In 1961, a simple and remarkable experiment revealed how this freakish change can be prevented and, almost uniquely in the vertebrates (animals with backbones), how human intervention can dramatically extend the lifespan of an organism far beyond its natural limits.

Cut out their gonads before the fish are fully developed, and kokanee salmon will live more than twice as long.[4] In a 1961 experiment, the last castrated male kokanee survived to age 8 years, over double the lifespan of their non-neutered controls.[5] Curiously, when all the control males died in their third year, so did "one of the castrated males in which a small piece of regenerated, testicular tissue was found at autopsy."[6] On average, the removed testicles weighed just 59 mg in total, the weight of a single grain of rice, but even a tiny scrap of unremoved testicle was still enough to cause catastrophic organ failure and death. The female benefit of neutering was only marginally smaller: the last of the non-neutered female control fish died later at age 4 years and 8 months, and the last neutered female died at

8.5 years.[7] The castrated super-ager fish would probably have survived even longer, but delightfully, a gang of raccoons broke in and ate many of the test subjects.

We can make kokanee double their normal limits of survival with scissors and stitches. It is a simple physical intervention that generates a huge extension of lifespan without altering telomeres, lowering oxidative damage rates, or replacing organs like an auto workshop. I find this intervention remarkable, and it has been replicated in this and many other salmonid species, but aging research seems to have forgotten the lesson. There was a brief flurry of discussion in the late 1960s, and there are occasional discussions of salmon's limited lifespan, but the idea now appears to be very obscure trivia.[8] There does not seem to be a tech bro, immortality-pill salesman, or singularity fanboy out there who seriously (or ever) considers the repercussions of salmon castration. Maybe they don't read Canadian fisheries journals, or maybe castration is just too difficult a sales pitch.

That's unfortunate because we can learn even more about survival if we dip our toes in other seas. Atlantic salmon, which again have a wonderful scientific name of *Salmo salar*, obviously live in the Atlantic. They do not have rivers long enough to compete with the Yukon, but they do speed up the Eastern Seaboard US and Canadian rivers, going hundreds of kilometers up the St. Lawrence, or 450 kilometers up the rivers from the Baltic into Finland. It seems they even used to run up the Hudson in New York before humans ruined everything. Like the Pacific salmon, Atlantic salmon endure the waterfalls, the shock transition to fresh water, and only a very occasional predator.

Atlantic salmon do something their Pacific lookalikes cannot. They can turn around and go back to the sea, and with enough luck they can do it again, and again, and again. These sea-returning salmon are common enough to have special names: fish that spawn and return to the sea are called kelts, and those that swim up the river and return without spawning are called baggots* for females and rawners for males. Kelts are common

---

*The oddly named baggots also count against the idea that salmon merely die from tiredness or shock. The typically younger and smaller baggots invest their energy into making eggs and swim upstream, wait for a month or two, and go back to the sea. Baggots survive in much better condition than kelts and outlive all the regular spawning salmon. I am guessing, but I really struggle to see how the act of scooping out a nest and ejecting eggs would require a fatally larger amount of energy, compared to the baggots' 900 km swim and the switch back to salt water.

but not the rule: most Atlantic salmon still suffer the same pattern as the Pacific salmon—they breed then die—and only about 10 percent of Atlantic salmon will return to the sea as kelts, baggots, or rawners.

This pattern really puts to bed the idea that salmon just sort of sadly die from exhaustion. In contrast with the lived experience of human parents like me, having kids does not mean you can just die from being tired. So why have kelts, which do not die after spawning, not become the only, or dominant, way of salmon life? Why would you *not* breed multiple times instead of giving up on life as soon as you spawn?

Salmon are clearly killed by sex, and it seems to be a trait that some members of a population can escape from. Yet this is not as singular an oddity as it seems. As we will see, sex-induced death, an evolved pattern of life called semelparity, is strangely common in the natural world. Our existence as a society even relies on several species that die just after they have sex. But before we learn about these species, we should go exploring some unusual cases to provide us with a better picture of sex-limited survival.

When people picture marsupials, the default is the iconic kangaroo, but there are many other marsupials of smaller size and greater diversity crawling and climbing and hopping across Australia. A group of such furry snufflers is the *Antechinus*: pronounced "anti-kye-ness" with a long Aussie drawl, if you were curious. It is also called the marsupial mouse by the imperial types and the *mardo* or maybe the *ngalugan* in the Wiradjuri language where I grew up. Picture a tiny shrew thing with a pouch.

Take the brown *Antechinus*, called *Antechinus stuartii*. These little blokes do something funny. They scurry out of their mother's nest and into the bush before, when the midwinter solstice starts, they begin collecting with the ladies into communal nests. Twenty or so *Antechinus* will live in a commune together. Then nature, as is often the case, puts humans to shame in a display of orgiastic sexuality. Corticosteroids skyrocket, and for two weeks or so the ten-month-old brown *Antechinus* males do nothing but—ahem—poke their proverbial into the unspeakable.

The marsupial mice travel from commune to commune, breeding with whomever they like (and often everyone) in a twenty-strong nest, multiple times a day.[9] For male *Antechinus*, sex hormone production overburdens their entire bodies: they go so hard at reproduction that their organs dissolve. Their limbs fall off. In the end, they are little more than a bag of barely breathing mouse jelly with an erection, and "synchronous immune system collapse and death quickly follow."[10] Scientists have dubbed this "suicidal reproduction"[11] or much more adroitly, the "Big Bang reproduction strategy."[12]

The best part of this is that the females are, despite the enthusiastic "cuddling," pretty much fine. They give a post-orgy shrug and scurry off into the undergrowth to enjoy a quiet month as a species without a single living male (outside a womb). That is, there is a big gap between the salmon-like male *Antechinus*, which copulate until their tails fall off, and the regular mouse-like females that often survive to breed for one or two more seasons.[13]

Two things can artificially alter when the male *Antechinus* breed and die. The first is the use of carefully timed artificial sunlight to "move" the winter solstice. For example, use some timers and sunlamps to fake a winter solstice two months early, and the poor little buggers will fire up their testicles two months early, breed with precisely nobody (the women aren't interested), and drop dead from the sex hormone overload anyway.[14] An orgy-free and sad existence. Lifespan extension also works: if you move the solstice two months later, the males will die two months late.[15]

The second intervention to change an *Antechinus* lifespan limit is the salmon method: cut out their testicles, and the males keep living far beyond the normal eleven months, even surviving up to two-and-a-half years.[16] Again, I don't see the lifespan extension community lining up to sell this stuff. Yet the little marsupials are fascinating because they can have their normal lifespan limits extended by simple interventions that have nothing to do with diet or telomeres, and that parallel the interventions that work in very different and independently evolving species like salmon. They also give us a better window on what kills them because the females completely avoid sex-death despite sharing the same environment, diet, and (except for the Y chromosomes) the same genes.*

---

*I can make an educated guess why. It takes exactly one male and one female to make one *Antechinus* baby, but only mum raises the kids. Parental care—the care of immediate kin—selects for conditions where the female hormones, ramped up during the breeding season, must be ramped back down again. In the males, however, there seems to be no selective advantage to the "off switch" of sex hormones: in closely related species the males disperse after sex, do not contribute to raising young, and have a very low chance of reaching a second breeding season. Without meaningful kin selection, there would be no selective advantage to a genetic program that turns the sex hormones off again. As such, if *Antechinus* males have lost the genetic program that says "breeding is over, stop increasing your production of sex hormones," what would the consequences be? Pretty much exactly what we observe:

That's fun, but we can cast our biological net even wider. We want to build up a little background knowledge before we reveal the amazing benefits of understanding the sex-death problem. Let's go to a forest and look at the bamboo.

Chinese bamboo* used to have another lyrical scientific name, *Bambusa bambusoides* (bamboo-sa bamboo-soy-deez), but traded half of it for the clunky *Phyllostachis bambusoides* (phil-oh-stack-iss). Whatever you call it, this bamboo will sit on a mountainside in Gifu or Hebei and quietly clone itself for 120 years. It will send off little shoots, expanding sideways to take over a mountainside, getting bigger and bigger and bigger.

Then, after 120 years of happy clone reproduction, the forest wakes up one day and chooses sex-death. The entire forest flowers, seeds explode from the seed pods, and the entire forest dies together: from the oldest shaft of timber in the canopy to the youngest-sprouting clone on the forest floor.[17]

Bamboo sex-death is not coordinated by some local signal through fungal networks or volatile gas-based signals spread by the plants. It is not governed by local conditions like rainfall either. We know this for a delightful reason. Chinese bamboo clones were transplanted to Japan before 1720 AD, using clones that flower across three different years. These "Japanese" clones were then reexported to the wider world from around 1880 onward, surviving in botanical collections worldwide. In 1966–1969, these clones did something amazing when "transplanted stocks of this bamboo in England, Alabama, and Russia, as well as their parental Japanese stock, flowered again" and died.[18] Clonal plants, separated by continents and living in utterly different places and climates, all flowered and died.

Exactly what causes this coordinated timing is unknown, but it transcended continents. The trigger of bamboo sex-death cannot be strictly

---

uncontrolled and fatal hormonal chaos. I would guess the *Antechinus* males are in a selective trap, where the ability to downregulate sex hormones has been lost, because the program to do so has near-zero odds of ever re-evolving—especially given that it may involve dozens of genes.

*The common name is now either Chinese or Japanese bamboo, according to your geopolitical and historical preference, because the bamboo was exported to Japan in antiquity, before these Chinese-origin clones were taken abroad from Japan. The bamboo may or may not have grown naturally in Japan before this, but the primary clone planted in botanic gardens seems to have come from China via Japan, so I have gone with "Chinese bamboo" out of biogeographical pedantry. Avoiding such choices is one good reason that we have scientific names.

seasonal because the bamboos see 480 seasons before death, but it does seem to happen at a tightly regulated time of the year. The bamboo clones must therefore carry an accurate internal death clock, one "set by an internal physiological calendar rather than an external weather cue," that keeps track of time while counting down to death.[19] Even young clones, made of cells that are a few weeks old at most, obey the death clock.

So why evolve such a thing? Why die on a schedule instead of just happily continuing to clone yourself, like seagrasses do for thousands of years, or avoid the post sex death and just flower repeatedly like the other plants?[20] If you somehow need to die from flowering, why not do it whenever the year and weather are optimal, or your size is at a maximum (for the littlest clones), rather than follow a fixed clock?

It is a superb mystery that, unlike many other aging research questions, is being slowly debated in a healthy research ecosystem.[21] I have no doubt that it will eventually be solved.

Bamboo is an important grass for humans. It is used in construction as "green steel" and in the making of very many culturally important, eco-friendly, and renewable products: traditional flutes, tatami mats, bamboo-cloth clothing, and the gorgeous bamboo screens of East Asian households. It absorbs massive amounts of carbon, triple the average forest, and is so strong and light it can even be used to make bicycle frames. However, these uses are of tiny importance compared to other grasses that suffer sex-death.

One out of every five calories consumed by humanity starts inside a grain of rice, a grain mostly eaten by the poor. Another 17–20 percent of human-consumed calories comes from wheat, which is mostly eaten by the rich but also sustains huge less well-off populations in the Middle East and North Africa. Yet another 5 percent of our calories comes from maize or corn,* including the large fraction of maize fed to animals that we then eat. Without these three species of plants, our societies would utterly collapse and most of humanity would die.

* Which has perhaps the most pleasant scientific name of all, *Zea mays* (zee-ya mays), in close competition with the black howler monkey *Alouatta caraya* (aloo-atta car-aye-yah, pronounced fast and loud and hooty like a howler monkey), and a microscopic pond monster named *Chaos* that never remains the same shape.

These grass species also follow a bamboo, salmon, and male *Antechinus* type of lifestyle—semelparity, or sex-death as I keep calling it. Although they skip the slowly-cloning-yourself stage seen in the bamboos, these crops synchronize their flowering, fill their grain babies with nutrients, dry out, and die less than a year after planting. A straight line from sex to grain filling "pregnancy" to death.

This time, however, we know how the death clock works.

For the first time in the book, I'm not going to pull this rug out from under you. We really understand it. To explain the system, I usually use a bicycle metaphor. It's a little silly, but it has seemed to help me before.

Imagine the wheat plant as a bicycle, cycling toward a wall of death. What does the journey look like? Well at first, the bicycle is "locked" when the wheat is just a seed. Only the correct conditions will unlock the plant bicycle: it is a combination lock that pops open for the right moisture, temperature, and time of year. This is why some seeds are hard to germinate: they have a very specific environment that unlocks germination, like *Eucalyptus* seeds that only germinate when they smell bushfire smoke, or the many types of seeds that germinate after being eaten and defecated (a process called endozoochory), like native North American blackberries that germinate if pooped out by a bird, and huckleberries that germinate at 140 times the normal rate if pooped out by a bear.[22]

Once unlocked, the plant bicycle will begin to move forward, pedaling past developmental milestones along the way, until it hits the wall of death. The most important milestones after germination (getting unlocked) are emergence (when the plant comes above the soil), the long "vegetative phase" where the plant sits around as a clump of floppy grass, "tillering" where a main stem or stems shoot up from the grass clump and make a spike with developing flowers on the end,* and "anthesis" when the plant begins flowering and becomes fertilized. Finally, the last three stages after fertilization all run rapidly together as the plant begins dying while filling up each grain with nutrients and energy from its own rapidly drying-up corpse. At the last of the milestones, the plant bicycle then strikes the wall of death as it drip-feeds the last bit of nutrients into a new generation of seed babies.

---

*If you are confused about grass or wheat flowers, it is because we don't usually see them: they hide inside a little capsule formed by two leaf-like structures, called the palea and the lemma, that fold around the little flowers like an envelope.

Once a plant is unlocked by germination, there are two processes that determine how fast it speeds past these milestones toward death: environmental inputs, which press on the pedals in our bicycle metaphor, and genetic inputs, which are the gears.

Light is the environmental input pushing hardest on the pedals: like the *Antechinus*, the plants pay very close attention to light. They will carefully measure and respond to the length of time they are kept in the dark, and whether the night is becoming longer or shorter, because it helps them measure the time of year.* The second-hardest pedal-pushing environmental force is temperature, with higher temperatures pushing the plant through development and toward death much faster. Finally, in a distant third place is water stress. Water stress pushes on the pedals but only under some circumstances, as the plant is better off breeding fast and making a few babies than running out of water and dying with none.

After these environmental pressures start pushing down on the pedals, several flowering time genes modify the rate at which the plant speeds through development. There are other genes in other plant species, such as the flowering time or "ft" genes in *Arabidopsis*, yet they fulfill the same basic gear-like function in regulating how fast you die. In wheat, for example, there are two primary gene families controlling the rate the plant pedals toward death: the front gear, constituting a gene called Photoperiod-1 or Ppd1, and back gears including the vernalization genes like Vrn2 and Vrn3.[23] In some varieties there are many different gear ratios and therefore a lot of speeds: a complex set of genotypes would be a thirty-speed bicycle, and a very simple genetic system would be a one-speed.

The upshot is that we can modify plant longevity to an extraordinary degree. We can change the gears on the bicycle through traditional plant breeding, the kind that has been practiced for a very long time and that you can do with very sharp scissors and a cotton swab. We also have newer methods that swap genes deliberately, but mostly it's still done by hand the old-fashioned way.

---

*Measuring the length of the nighttime also allows plants to set aside food, that they make during the day, as a kind of starchy midnight snack so they do not go hungry when they cannot photosynthesize. If you artificially make the night longer, plants will run out of their starch-granule lunchbox and begin to starve, digesting important parts of themselves as an emergency measure.

Even old varieties of wheat have routinely been selected to have very different lifespans. "Winter" wheats will flower after about six to eight months, after sitting dormant over winter (when temperature is not pushing on the pedals) in their vegetative "sit still and do nothing" phase. In contrast, "spring" wheats—which may only differ from winter wheats by a single gene across their massive sixteen-billion-letter genome—can develop and die after as little as three months. My old boss, James (Ben) Trevaskis, even had recombinant inbred lines of wheat that only differed by a single gene, which changed the "bicycle gear," in their entire genome. Genetic lifespan extension, sitting in his desk drawer in a manila envelope.

We are absolute bloody masters at lifespan modification in plants, to the immense benefit of society.

We can also modify plant lifespans with environmental changes: some as simple as a thermostat or a sunlamp. You can even do it yourself.

As a little background, these crop plants have tiny hidden flowers pollinated by the wind or by themselves. The ovules and pollen of these flowers are very temperature sensitive: if the developing flowers suffer a frost, they will die; if the weather is too cold and damp, they will suffer fungal diseases; and if the weather is too warm, the pollen will become sterile and fail to fertilize the female flower parts. They are the Goldilocks's porridge of sex. Not too hot and not too cold, juuuust right. If we want to have a good crop, we have to plant at the right time and have a seed that takes the correct amount of time to develop so that it flowers just after the frost and just before the heat.

The annual timing of "when is juuuust right" varies enormously across the grain-growing regions of the world. When we move plants to a new latitude or region, the best time for flowering also needs change to fit inside the best temperature window for a good crop. Yet flowering in these plants causes death. The outcome is that plant scientists are constantly manipulating plant lifespans to get the right flowering time, to an extent that puts lifespan extension gurus to absolute shame.

It isn't just the grasses that we understand or rely on when we manipulate lifespan either. Tomato lifespan is coregulated by a couple of "branching panicle" genes that govern flower arrangements and clustering. Change these genes, and the flowers will die faster or slower and so will the plant. Or take tulips: you can regulate when they breed with some midnight LED lights. Trick them into thinking the nights are spring length, and they will flower when you prefer. Then they will die faster.

One of my coauthors helped boost the popularity of "speed breeding": manipulating light conditions to grow as many generations of plants as rapidly as possible, thereby ramping up the speed of evolution and artificial selection for plant scientists in developing countries. All he did was change the lighting so that the plant was drenched in a constant and very specific wavelength of disco purple for twenty-four hours a day. Wheat lifespans can be hugely compressed using this method, allowing us to grow six generations a year in a shipping container rather than one to two generations in the field.[24] This is like accelerating the human lifespan so much that you leave school at age six, start having kids at seven, retire at twenty-two, and die in old age at twenty-six.

Improving our prediction of when plants flower, and therefore die, was my job for three years as a government scientist—where I worked with industrial giants like Bayer, BASF, and Limagrain, which routinely modify the lifespan of plants. I was focused on using the modern surrogate for critical thinking, machine learning, to interrogate satellite, field, and genomic data and predict complex traits like longevity.[25] But altering plant lifespans is not necessarily a complex process requiring a supercomputer. Storing living seeds or tubers for as long as we need them—environment-driven lifespan extension, if you like—is the basis of our entire society. Humans have been doing it incidentally or on purpose since the bronze age.

You can do it, right now, very easily.

Take a seed, like a whole date. Place it in a drawer. Congratulations. A date seed, wrapped in a date, is like a dormant child in a spaceship. It will go into static shutdown mode, slowly consuming its packet of energy just to maintain and repair itself, until conditions are right. Slowly adjusting its lifespan to match the environment—and since we control the environment, we control the longevity of the little date seed. The extent of this delay can be remarkable. Around 1963–1965, a cluster of date seeds were discovered in the foundations of a fortress built by Herod, beneath the desert of Judea, and carbon-dated to be almost two thousand years old.[26] In 2005 they were given the right conditions, and a hormonal helper, and began to grow.[27] They were alive, without water or light, being slowly irradiated beneath one of the harshest deserts on earth for two thousand years. Patiently waiting, as the toughest organism alive.

I find this kind of lifespan extension amazing, even when people do it every day. Yet the most stupendous, world-changing modification of plant lifespan may be the news from southern China and the Philippines.

Rice feeds 2.5 billion of the world's poorest people. It is also a shockingly difficult thing to sow and harvest: work that is largely carried out by hundreds of millions of women and girls, who miss school to sow the crops and break their backs for every bloody bowl. They do this twice or three times a year, and have done for eleven thousand years, because rice suffers sex-death and must be replanted.

Or I should say, they *used to have to do it.*

In a triumph of understanding and ingenuity, scientists from Yunnan and the Philippines have done the near-unthinkable. They made rice that has eight lives.[28] You plant it, the rice grows and flowers and produces a crop, and the stems and panicles die. The roots don't die. They come back up, without replanting, seven more times. For seven more crops.

Without the physical cost of sowing or the need to regrow their roots, these perennial rice crops require half of the labor input and less than half the pesticide use. They need less fertilizer, which saves poor people money, and helps keep the waterways clean. Not having to plow the fields twice a year prevents massive amounts of soil erosion and puts less nitrogen into the rivers because you use less fertilizer.[29] When the varieties derived from this breakthrough are sent out, hundreds of millions of women will no longer skip school to slog through the heavy mud of the paddy, the poor will pocket the money, and the environment will clean up.

The project was started by the chronically underfunded International Rice Research Institute (IRRI), which sits like a James Bond villain's lair on the flanks of a volcano, filled with '50s Americana-style houses, unnecessarily guarded by razor wire and a bloke with an assault rifle. The IRRI project was cut after budget cuts because the world is a shitty short-sighted place, but the project was rescued when the lead author Hu Fengyi brought his project to the Yunnan Academy of Agricultural Sciences. On a tiny budget, and with the help of plant scientists who wish to feed the poor, the brilliant Fengyi and his team outperformed and thoroughly outclassed every charlatan in aging science. They intentionally increased rice lifespan 800 percent and did something that may change the face of society.

The vanity-driven bullshit of Calico labs looks pale.

It's a masterpiece. It is also just a beginning. If we can do this in rice, we may yet do it in wheat, in maize, in every plant that needs to be plowed into the fields every year,* at great monetary expense and great physical cost to the soil and the water and the air.

I do not care to have an immortal, bilious billionaire, but I cannot wait for perennial wheat, maize, or oats.[30]

Nonaging biology has made strides in our understanding of lifespan and aging because short-term cure-me-now funding has been generally absent. Curiosity and time, not money and salesmanship, have allowed us to extended organic life.

What aging research can learn from these advances is simple: sack the charlatans, laugh out anyone who promises a "cure" to aging or carries a trademark, and give space to reproducible basic research. Another aging researcher obsessed with pilfering billionaires, monetizing hope, and medicalizing the well won't help anyone.

Yet that does not make the case for aging research hopeless. It just means that new advances will not come from inside aging research—or longevity science, healthspan research, or any such new-name woo—but from traditional slow-paced blue-sky biology.

Perennial rice is the product of a healthy research ecosystem, the absence of fraudulent wankers, and the capacity to perform meaningful experiments with minimal hype. It may not live up to its full promise, but whatever happens, it is a remarkable achievement. It is one of many in the quiet parts of biology, like the salmon-neutering and the *Antechinus* research, where success is achieved via thought and care.

The contrast with aging research could not be more stark.

The big-bang reproduction of bad ideas continues apace in aging research, making great leaps of reasoning to swim short distances, before the fervent mating of money and fishy science ejaculates cash onto a shallow rocky

---

*It is not clear how this would work when you have interspecies multiple cropping—like the alternating rice-wheat rotations I've seen in South Korea or northern China. Yet it's a brave new world. Perhaps you can do interrow perennial crop plantings so that the wheat and rice pop up out of the ground in alternating cycles. With precision planting, anything is possible.

bed of bad eggs. The progenitors of these once-great ideas will eventually die and wash downstream, unnoticed, returning to where they started in a great and meaningless circle.

The best I can hope for is to hunch over like a solitary bear, catching the highest leaps of logic at the waterfall, and crunch up some hook-jawed hopefuls with my mouth. Like the bears, I doubt this will make much difference. The salmon will be back next year.

Acknowledgments

Many people have helped me with the book through friendship and support, and many, many others have unhelped me to an extraordinary degree: placing obstacles to all rational progress and conflict resolution in my path, to the point that I simply have no alternative but to ball up my questions into a self-effacing comedy and place it in a book, in a doomed attempt to overcome their hindrance.

One person helped me above all others: Dr. Elena Racheva, who was my personal superhero. Her notes were thoughtful, her company brilliant, and I shall not forget it.

Notes

Preface

1. Phillips, C. “The Golden State Killer Investigation and the Nascent Field of Forensic Genealogy.” *Forensic Science International: Genetics* 36, 186–188 (2018).

2. How-Kit, A., Sahbatou, M., Hardy, L. M., et al. “The CEPH Aging Cohort and Biobank: A Valuable Collection of Biological Samples from Exceptionally Long-Lived French Individuals and Their Offspring for Longevity Studies.” *Geroscience* 46, 2681–2695 (2024).

3. Hoad, P. “‘People Are Caught Up in Magical Thinking’: Was the Oldest Woman in the World a Fraud?” *The Guardian*. November 30, 2019. https://www.theguardian.com/science/2019/nov/30/oldest-woman-in-the-world-magical-thinking.

4. Hoad, “‘People Are Caught Up in Magical Thinking.’”

5. Hoad, “‘People Are Caught Up in Magical Thinking.’”

Chapter 1

1. Dong, X., Milholland, B. & Vijg, J. “Evidence for a Limit to Human Lifespan.” *Nature* 538, 257–259 (2016).

2. Newman, S. J. & Furbank, R. T. “Explainable Machine Learning Models of Major Crop Traits from Satellite-Monitored Continent-Wide Field Trial Data.” *Nature Plants* (2021).

3. Newman & Furbank. “Explainable Machine Learning Models of Major Crop Traits.”

4. Dong, Milholland & Vijg. “Evidence for a Limit to Human Lifespan.”

5. Dong, Milholland & Vijg. “Evidence for a Limit to Human Lifespan.”

6. Newman, S. J. & Easteal, S. “The Dynamic Upper Limit of Human Lifespan.” *F1000Res* **6** (2017).

7. Newman & Easteal. "The Dynamic Upper Limit of Human Lifespan."

8. Newman & Easteal. "The Dynamic Upper Limit of Human Lifespan."

9. Dong, Milholland & Vijg. "Evidence for a Limit to Human Lifespan."

10. Newman & Easteal. "The Dynamic Upper Limit of Human Lifespan."

11. "Top 100 Articles—2016." Altmetric. https://www.altmetric.com/top100/2016/.

12. Newman & Easteal. "The Dynamic Upper Limit of Human Lifespan."

13. Brown, N. J. L., Albers, C. J. & Ritchie, S. J. "Contesting the Evidence for Limited Human Lifespan." *Nature* 546, E6–E7 (2017); Lenart, A. & Vaupel, J. W. "Questionable Evidence for a Limit to Human Lifespan." *Nature* 546, E13–E14 (2017); Rozing, M. P., Kirkwood, T. B. L. & Westendorp, R. G. J. "Is There Evidence for a Limit to Human Lifespan?" *Nature 2017* 546, E11–E12 (2017); Hughes, B. G. & Hekimi, S. "Many Possible Maximum Lifespan Trajectories." *Nature* 546, E8–E9 (2017); De Beer, J., Bardoutsos, A. & Janssen, F. "Maximum Human Lifespan May Increase to 125 Years." *Nature* 546, E16–E17 (2017).

14. "Peer Review Post-Mortem: How a Flawed Aging Study Was Published in Nature." Showcase. https://showcase.casw.org/stories/peer-review-post-mortem-how-a-flawed-aging-study-was-published-in-nature/.

15. "Peer Review Post-Mortem." Showcase.

16. "Peer Review Post-Mortem." Showcase.

17. "Peer Review Post-Mortem." Showcase.

18. "Review of Evidence for a Limit to Human Lifespan." Publons. https://publons.com/wos-op/review/480517/.

19. "Peer Review Post-Mortem." Showcase.

20. "Widely Publicized Nature Study on Human Age Limit Draws Fire." Retraction Watch (2017). https://retractionwatch.com/2017/06/28/widely-publicized-nature-study-human-age-limit-draws-fire/.

21. "Peer Review Post-Mortem." Showcase.

22. van Santen, H. "Nature Article Is Wrong About 115 Year Limit on Human Lifespan." *NRC* (2016).

23. Newman, S. J. "Errors as a Primary Cause of Late-Life Mortality Deceleration and Plateaus." *PLoS Biology* **16**, e2006776 (2018).

24. Newman. "Errors as a Primary Cause of Late-Life Mortality Deceleration and Plateaus."

25. Newman, "Errors as a Primary Cause of Late-Life Mortality Deceleration and Plateaus."

26. Vaupel, J. W., Manton, K. G. & Stallard, E. "The Impact of Heterogeneity in Individual Frailty on the Dynamics of Mortality." *Demography* 16, 439 (1979).

27. Newman, "Errors as a Primary Cause of Late-Life Mortality Deceleration and Plateaus."

28. Barbi, E., Lagona, F., Marsili, M., Vaupel, J. W. & Wachter, K. W. "The Plateau of Human Mortality: Demography of Longevity Pioneers." *Science* 360, 1459–1461 (2018).

29. Barbi, Lagona, Marsili, Vaupel & Wachter. "The Plateau of Human Mortality."

30. Barbi, Lagona, Marsili, Vaupel & Wachter. "The Plateau of Human Mortality."

31. Barbi, Lagona, Marsili, Vaupel & Wachter. "The Plateau of Human Mortality."

32. Newman, S. J. "Plane Inclinations: A Critique of Hypothesis and Model Choice in Barbi et al." *PLoS Biology* 16, e3000048 (2018).

33. Newman. "Plane Inclinations."

34. Newman. "Plane Inclinations."

35. Newman, S. J. "Unsupported Choices Generate a Plateau." *Science* (2018).

36. Newman. "Plane Inclinations."

37. Wachter, K. W. "Hypothetical Errors and Plateaus: A Response to Newman." *PLoS Biology* 16, e3000076 (2018).

38. Wachter. "Hypothetical Errors and Plateaus."

39. Wachter. "Hypothetical Errors and Plateaus."

Chapter 2

1. Sear, R., Mace, R., & McGregor, I. A. "Maternal grandmothers improve nutritional status and survival of children in rural Gambia." *Proceedings of the Royal Society B: Biological Sciences* 267, 1641–1647 (2000).

2. "De Die Natale." Censorinus. https://elfinspell.com/ClassicalTexts/Maude/Censorinus/DeDieNatale-Part1.html.

3. Censorinus. "De Die Natale."

4. Dong, X., Milholland, B. & Vijg, J. "Evidence for a Limit to Human Lifespan." *Nature* 538, 257–259 (2016).

5. Newman, S. J. & Easteal, S. "The Dynamic Upper Limit of Human Lifespan." *F1000Research* 6 (2017).

6. Newman & Easteal. "The Dynamic Upper Limit of Human Lifespan."

7. Newman & Easteal. "The Dynamic Upper Limit of Human Lifespan."

8. Newman & Easteal. "The Dynamic Upper Limit of Human Lifespan."

9. Newman & Easteal. "The Dynamic Upper Limit of Human Lifespan."

10. Newman, S. J. "Errors as a Primary Cause of Late-Life Mortality Deceleration and Plateaus." *PLoS Biology* 16, e2006776 (2018).

11. Vaupel, J. W., Manton, K. G. & Stallard, E. "The Impact of Heterogeneity in Individual Frailty on the Dynamics of Mortality." *Demography* 16, 439 (1979).

12. Horiuchi, S. & Wilmoth, J. R. "Deceleration in the Age Pattern of Mortality at Older Ages." *Demography* 35, 391–412 (1998); Modig, K., Andersson, T., Vaupel, J., Rau, R. & Ahlbom, A. "How Long Do Centenarians Survive? Life Expectancy and Maximum Lifespan." *The Journal of Internal Medicine* 282, 156–163 (2017).

13. Barbi, E., Lagona, F., Marsili, M., Vaupel, J. W. & Wachter, K. W. "The Plateau of Human Mortality: Demography of Longevity Pioneers." *Science* 360, 1459–1461 (2018); Horiuchi & Wilmoth. "Deceleration in the Age Pattern of Mortality at Older Ages"; Alvarez, J.-A., Villavicencio, F., Strozza, C. & Camarda, C. G. "Regularities in Human Mortality After Age 105." *PLoS One* 16, e0253940 (2021).

14. Pletcher, S. & Curtsinger, J. "Mortality Plateaus and the Evolution of Senescence: Why Are Old-Age Mortality Rates So Low?" *Evolution (NY)* 52, 454–464 (1998); Mueller, L. D. & Rose, M. R. "Evolutionary Theory Predicts Late-Life Mortality Plateaus." *Proceedings of the National Academy of Sciences* 93, 15249–15253 (1996); Mueller, L. D., Rauser, C. L. & Rose, M. R. *Does Aging Stop?* (Oxford University Press, Oxford, 2011). doi:10.1093/acprof:oso/9780199754229.001.0001; Pletcher & Curtsinger. "Mortality Plateaus and the Evolution of Senescence."

15. Zarulli, V. "Unobserved Heterogeneity of Frailty in the Analysis of Socioeconomic Differences in Health and Mortality." *European Journal of Population* (2016). doi:10.1007/s10680-015-9361-1; Manton, K. G. & Stallard, E. "Methods for Evaluating the Heterogeneity of Aging Processes in Human Populations Using Vital Statistics Data: Explaining the Black/White Mortality Crossover by a Model of Mortality Selection." *Human Biology* (1981).

16. Zarulli. "Unobserved Heterogeneity of Frailty in the Analysis of Socioeconomic Differences in Health and Mortality"; Manton & Stallard. "Methods for Evaluating the Heterogeneity of Aging Processes."

17. Newman & Easteal. "The Dynamic Upper Limit of Human Lifespan."

18. Wachter, K. W. "Hypothetical Errors and Plateaus: A Response to Newman." *PLoS Biology* 16, e3000076 (2018).

19. Mueller, Rauser & Rose. *Does Aging Stop?*

Chapter 3

1. October Films Production. *The Man Who Would Be Kubrick: Alan Conway Interview* (Channel Four Television Corporation, London, 1999).

2. Anthony, A. "The Counterfeit Kubrick." *The Guardian* (1999). https://www.theguardian.com/film/1999/mar/14/andrewanthony.

3. October Films Production. *The Man Who Would Be Kubrick.*

4. October Films Production. *The Man Who Would Be Kubrick.*

5. Anthony. "The Counterfeit Kubrick."

6. "Gerontology Research Group Supercentenarian Index page." Gerontology Research Group (2016). http://www.grg.org/SC/SCindex.html; Robine, J.-M., Gampe, J. & Vaupel, J. "IDL, the International Database on Longevity." *Gerontology* 1–30 (2005).

7. Wilmoth, J. "Christian Mortensen, 1882–1998" (1998). https://u.demog.berkeley.edu/~jrw/tribute.htm.

8. Gondo, Y., Hirose, N., Yasumoto, S., Arai, Y. & Saito, Y. "Age Verification of the Longest Lived Man in the World." *Experimental Gerontology* (2017). doi:10.1016/j.exger.2017.08.030.

9. Gondo, Hirose, Yasumoto, Arai & Saito. "Age Verification of the Longest Lived Man in the World."

10. Gondo, Hirose, Yasumoto, Arai & Saito. "Age Verification of the Longest Lived Man in the World."

11. Gondo, Hirose, Yasumoto, Arai & Saito. "Age Verification of the Longest Lived Man in the World."

12. Gondo, Hirose, Yasumoto, Arai & Saito. "Age Verification of the Longest Lived Man in the World."

13. How-Kit, A., Sahbatou, M., Hardy, L. M., et al. "The CEPH Aging Cohort and Biobank: A Valuable Collection of Biological Samples from Exceptionally Long-Lived French Individuals and Their Offspring for Longevity Studies." *Geroscience* 46, 2681–2695 (2024).

14. Ohnuki-Tierney, E. *Kamikaze Diaries: Reflections of Japanese Student Soldiers* (University of Chicago Press, Chicago, 2006).

15. Ohnuki-Tierney. *Kamikaze Diaries.*

16. Fuchs, E., Kasahara, T., Saaler, S. *A New Modern History of East Asia* (Vandenhoeck & Ruprecht Gmbh & Co., Göttingen, Germany, 2017); Yeonok, K. "Reexamine of

the 1920 Massacre of Norapawie (JangAm) Villages at the Kando." *Ilbon-Hak (Journal of Japanology)* 52, 117–142 (2020).

17. Gondo, Hirose, Yasumoto, Arai & Saito. "Age Verification of the Longest Lived Man in the World."

18. Gondo, Hirose, Yasumoto, Arai & Saito. "Age Verification of the Longest Lived Man in the World."

19. Gondo, Hirose, Yasumoto, Arai & Saito. "Age Verification of the Longest Lived Man in the World."

20. Gondo, Hirose, Yasumoto, Arai & Saito. "Age Verification of the Longest Lived Man in the World."

21. Gondo, Hirose, Yasumoto, Arai & Saito. "Age Verification of the Longest Lived Man in the World."

22. Young, R. "Age 115 or More in the United States: Fact or Fiction?" in *Supercentenarians* (Springer, Berlin, 2010). doi:10.1007/978-3-642-11520-2_15.

23. O'Connor v. Donaldson: 422 U.S. 563 (1975): Justia US Supreme Court Center. https://supreme.justia.com/cases/federal/us/422/563/; Donaldson, K. *Insanity Inside Out* (Crown Publishers, New York, 1976), 338.

24. "Kenneth Donaldson Discusses His Book 'Insanity Inside Out.'" Studs Terkel Radio. https://studsterkel.wfmt.com/programs/kenneth-donaldson-discusses-his-book-insanity-inside-out.

25. "To Carrie: Happy Birthday, Sweet 115. The Oldest American Ever Spent 75 years in Chattahoochee." *Orlando Sentinel* (1989).

26. "Chronicles from Chattahoochee." Sally J. Ling. https://sallyjling.com/chronicles-chattahoochee/.

27. "Carrie White, 116; Listed as World's Oldest Person." *Los Angeles Times* (1991). https://www.latimes.com/archives/la-xpm-1991-02-15-mn-1246-story.html; *Orlando Sentinel*. "To Carrie: Happy Birthday, Sweet 115."

28. "At 116, She Has Her Cake and Honors, Too." *Los Angeles Times* (1990). https://www.latimes.com/archives/la-xpm-1990-11-19-mn-3586-story.html; *Los Angeles Times*. "Carrie White, 116."

29. "Juan Vicente Pérez es ahora el hombre más longevo del mundo." Diario La Nación. https://lanacionweb.com/destacado/juan-vicente-perez-es-ahora-el-hombre-mas-longevo-del-mundo/#google_vignette.

30. Central Intelligence Agency. *CIA World Factbook: Life Expectancy at Birth* (CIA, Washington, DC, 2022).

31. Levin, J. "Oldest American Is 113-Year-Old Jersey Girl." *USA Today* (2016). https://www.usatoday.com/story/news/nation-now/2016/07/21/oldest-american-year-old-jersey-girl/87373424/.

32. Young, R. D. "African American Longevity Advantage: Myth or Reality? A Racial Comparison of Supercentenarian Data." PhD thesis (Georgia State University, 2008). doi: https://doi.org/10.57709/1062194.

33. Brown, J. "America's Oldest Living WWII Veteran Turns 112, Fueled by Cigars, Dr Pepper and Coffee." Business Insider (2018). https://www.businessinsider.com/oldest-person-in-america-us-world-war-ii-veteran-age-richard-overton-2018-5; Christie, J. "The Oldest Living US Veteran, Age 108, Credits Drinking Whiskey and Smoking a Dozen Cigars Every Day for His Long Life." *Daily Mail* (2014). https://www.dailymail.co.uk/news/article-2830888/The-oldest-living-veteran-smokes-drinks-whiskey-day-drives-does-yard-work-s-108-YEARS-OLD.html.

34. "Texas Vital Statistics Indexes." Texas State Library and Archives Commission (2021). https://www.tsl.texas.gov/arc/vitalfaq.

35. Brumberg, H. L., Dozor, D. & Golombek, S. G. "History of the Birth Certificate: From Inception to the Future of Electronic Data." *Journal of Perinatology* 32, 407–411 (2012); Preston, S. H., Elo, I., Rosenwaike, I. & Hill, M. "African-American Mortality at Older Ages." *Demography* 33, 193–209 (1996).

36. US Census Bureau. *Thirteenth Census of the United States: 1910* (US Census Bureau, Washington, DC, 1913).

37. "Richard Arvin Overton (1906–2018)." FamilySearch. https://www.familysearch.org/tree/person/sources/GMKP-44V; US Census Bureau. *Fourteenth Census of the United States: 1920* (US Census Bureau, Washington, DC, 1922).

38. FamilySearch. "Richard Arvin Overton (1906–2018)."

39. FamilySearch. "Richard Arvin Overton (1906–2018)."

40. Brown. "America's Oldest Living WWII Veteran."

41. "Pensioni: nascondeva nel congelatore la madre (103 anni) morta da 6 mesi. E ritirava l'assegno." Blitz Quotidiano. https://www.blitzquotidiano.it/cronaca-italia/pensioni-nascondeva-nel-congelatore-la-madre-103-anni-morta-da-6-mesi-e-ritirava-lassegno-2541814/.

42. "Biella, Corpse of the Mother Hidden in the Freezer: Her Son Denounced." Today (2016). https://www.today.it/cronaca/nasconde-cadavere-congelatore-biella.html.

43. "Pensioni." Blitz Quotidiano.

44. "Pensioni." Blitz Quotidiano.

45. "Pensioner's Mummified Body Found in Wall." The Local (2013). https://www.thelocal.it/20131125/italian-pensioners-mummified-body-found-in-wall; The Local. "Pensioner's Mummified Body Found in Wall"; "'Your Mum's 104?': Man Caught in Pension Scam." The Local (2014). https://www.thelocal.it/20140407/benefit-fraudster-nabbed-by-mums-longevity-2.

46. "Son Kept Dead Mum in Freezer for Her Pension." The Local (2014). https://www.thelocal.it/20140217/son-kept-dead-mum-in-freezer-for-her-pension.

47. "Son 'Kept His Dead Mother in a Freezer So He Could Keep Collecting Her Pension for More Than Six Months.'" Daily Mail Online (2013). https://www.dailymail.co.uk/news/article-2519800/Son-kept-dead-mother-freezer-collecting-pension-months.html; O'Leary, A. "Man Kept Dead Mum's Corpse in Freezer for 6 Months to Continue Collecting Pension." Mirror Online (2020). https://www.mirror.co.uk/news/world-news/man-kept-dead-mums-corpse-22449483.

48. "Carlow: Two Jailed After Using Body to Claim Dead Man's Pension." BBC News (2023). https://www.bbc.co.uk/news/articles/c1d7gvlxerdo.

49. McCárthaigh, S. & McCullough, M. "Woman Claimed Dead Father-In-Law's Pension for 30 Years and Got Husband to Pretend to Be Him." Belfast Live (2024). https://www.belfastlive.co.uk/news/woman-claimed-dead-father-laws-29357495.

50. McCárthaigh & McCullough. "Woman Claimed Dead Father-In-Law's Pension for 30 Years."

51. Desjardins, B. "Validation of Extreme Longevity Cases in the Past: The French-Canadian Experience" in *Validation of Exceptional Longevity* (Odense University Press, Odense, 2003).

52. Desjardins. "Validation of Extreme Longevity Cases in the Past."

53. Desjardins. "Validation of Extreme Longevity Cases in the Past."

54. McWhirter, N. *Guinness Book of World Records* (Sterling Pub, New York, 1980), 672; "Charlie Smith, Ex-Slave, Is Dead; Believed Oldest American at 137." *New York Times* (1979). https://www.nytimes.com/1979/10/07/archives/charlie-smith-exslave-is-dead-believed-oldest-american-at-137-i.html.

55. *New York Times*. "Charlie Smith, Ex-Slave, Is Dead"; Belcher, W. & Heery, B. "Is Charlie Smith 136 Years Old? Nobody Knows." *Tampa Tribune* (1979).

56. "Interview with Charlie Smith, Bartow, Florida, March 17, 1975." American Memory from the Library of Congress. https://web.archive.org/web/20041030152251/http://memory.loc.gov/cgi-bin/query/r?ammem/afcesn:@field(DOCID+afc9999001t17510).

57. American Memory from the Library of Congress. "Interview with Charlie Smith."

58. Chaikin, A. *A Man on the Moon: The Voyages of the Apollo Astronauts* (Penguin Books, New York, 1994), 670.

59. *New York Times*. "Charlie Smith, Ex-Slave, Is Dead."

60. Coles, L. S. "Aging: The Reality: Demography of Human Supercentenarians." *The Journals of Gerontology, Series A* 59, B579–B586 (2004).

61. Maier, H., Gampe, J., Jeune, B., Robine, J.-M. & Vaupel, J. W. "Supercentenarians." *Demographic Research Monographs* 7 (2010).

62. Associated Press. "Adele Dunlap, the Oldest American Person, Dies Aged 114." *The Guardian* (2017).

63. Buncombe, A. "Agnes Fenton: Woman Who Drank Three Beers and a Shot of Whiskey Every Day for 70 Years Has Just Turned 110." *The Independent* (2015).

64. Smith, D. "RAF Flypast for Europe's Oldest Man." *The Guardian* (2008). https://www.theguardian.com/uk/2008/jun/01/military.firstworldwar.

65. Maier, Gampe, Jeune, Robine & Vaupel. "Supercentenarians."

66. Maier, Gampe, Jeune, Robine & Vaupel. "Supercentenarians."

67. Maier, Gampe, Jeune, Robine & Vaupel. "Supercentenarians."

68. "Oldest American Man, WWII Veteran to be Buried at Historic Cemetery." United States Army. https://www.army.mil/article/215936/oldest_american_man_wwii_veteran_to_be_buried_at_historic_cemetery.

Chapter 4

1. "Oldest Dog EVER Record Broken by 30-Year-Old Bobi from Portugal." Guinness World Records (2023). https://www.guinnessworldrecords.com/news/2023/2/oldest-dog-ever-record-broken-by-30-year-old-bobi-from-portugal-736224.

2. Guinness World Records. "Oldest Dog EVER Record Broken by 30-Year-Old Bobi from Portugal."

3. Reynolds, M. "A Scandal Is Tearing the World of Record-Breaking Dogs Apart." *Wired* (2024). https://www.wired.com/story/guinness-worlds-oldest-dog-scandal/.

4. Reynolds. "A Scandal Is Tearing the World of Record-Breaking Dogs Apart."

5. Callery, J. "Bobi 'the World's Oldest Dog' LOSES His Title—Four Months After He Died: Guinness World Records Say There Is 'No Conclusive Evidence' to Support Claim He Was 31 Years Old." Daily Mail Online (2024). https://www.dailymail.co.uk/news/article-13113451/Bobi-worlds-oldest-dog-LOSES-title-four-months-died-Guinness-World-Records-say-no-conclusive-evidence-support-claim-31-years-old.html.

6. Hall, R. "Claims About Genuine Age of Bobi, World's Oldest Dog, to Be Investigated." *The Guardian* (2023).

7. Reynolds, M. "A Scandal Is Tearing the World of Record-Breaking Dogs Apart."

8. Horner, J. R. & Padian, K. "Age and Growth Dynamics of Tyrannosaurus Rex." *Proceedings of the Royal Society of London. Series B, Biological Sciences* **271**, 1875–1880 (2004).

9. Gondo, Y., Hirose, N., Yasumoto, S., Arai, Y. & Saito, Y. "Age Verification of the Longest Lived Man in the World." *Experimental Gerontology* (2017). doi:10.1016/j.exger.2017.08.030.

10. Newman, S. J. "Supercentenarian and Remarkable Age Records Exhibit Patterns Indicative of Clerical Errors and Pension Fraud." *bioRxiv* (2020). doi: https://doi.org/10.1101/704080.

11. Timmer, M., Baten, J., d'Ercole M. M., Rijpma, A. & Timmer, M. P. *How Was Life?: Global Well-Being Since 1820* (OECD, Paris, 2014); Cappelli, G. & Vesta, M. "A 'Silent Revolution': School Reforms and Italy's Educational Gender Gap in the Liberal Age (1861–1921)." *Cliometrica* 15, 203–229 (2021).

12. Gondo, Hirose, Yasumoto, Arai & Saito. "Age Verification of the Longest Lived Man in the World."

13. "Tokyo's 'Oldest Man' Had Been Dead for 30 Years." BBC News. https://www.bbc.co.uk/news/world-asia-pacific-10809128.

14. "Mummy Believed to Be That of '111-Year-Old' Man Found in Tokyo." Japan Today. https://web.archive.org/web/20101213124104/http://japantoday.com/category/crime/view/mummy-believed-to-be-that-of-111-year-old-man-found-in-tokyo.

15. Japan Today. "Mummy Believed to Be That of '111-Year-Old' Man Found in Tokyo."

16. Japan Today. "Mummy Believed to Be That of '111-Year-Old' Man Found in Tokyo."

17. "Tokyo's 'Oldest Woman' Missing for Decades." BBC News. https://www.bbc.co.uk/news/world-asia-pacific-10848254.

18. "About Family Register Office Work to Affect Location Unknown Elderly People." Japanese Ministry of Justice (2010). http://www.moj.go.jp/MINJI/minji04_00008.html; Fackler, M. "Japan, Checking on Its Oldest, Finds Many Gone." *New York Times* (2010).

19. Japanese Ministry of Justice. "About Family Register Office Work to Affect Location Unknown Elderly People."

20. Japanese Ministry of Justice. "About Family Register Office Work to Affect Location Unknown Elderly People."

21. Kannisto, V. *Development of Oldest-Old Mortality, 1950–1990: Evidence from 28 Developed Countries* (Odense University Press, Odense, 1994), 108.

22. Willcox, D. C., Willcox, B. J., He, Q., Wang, N. C. & Suzuki, M. "They Really Are That Old: A Validation Study of Centenarian Prevalence in Okinawa." *Journals of Gerontology* (2008). doi:10.1093/gerona/63.4.338.

23. Saito, Y., Yong, V. & Robine, J.-M. "The Mystery of Japan's Missing Centenarians Explained." Demographic Research 26, 239–252 (2012). doi:10.4054/DemRes.2012.26.11.

24. Saito, Yong & Robine. "The Mystery of Japan's Missing Centenarians Explained."

25. Gondo, Hirose, Yasumoto, Arai & Saito. "Age Verification of the Longest Lived Man in the World"; Willcox, Willcox, He, Wang & Suzuki, M. "They Really Are That Old"; Gondo, Y., Hirose, N., Arai, Y., et al. "Functional Status of Centenarians in Tokyo, Japan: Developing Better Phenotypes of Exceptional Longevity." *Journals of Gerontology* 61 (2006). doi:10.1093/gerona/61.3.305.

26. Yang, S. S. & Huang, H. C. "The Impact of Longevity Risk on the Optimal Contribution Rate and Asset Allocation for Defined Contribution Pension Plans." *Geneva Papers on Risk and Insurance: Issues and Practice* 34, 660–681 (2009).

27. Kannisto. *Development of Oldest-Old Mortality*, 108; Vaupel, J. W., Carey, J. R., Christensen, K., et al. "Biodemographic Trajectories of Longevity." *Science* 280, 855–860 (1998). https://doi.org/10.1126/science.280.5365.855.

28. UN Population Division. "World Population Prospects: the 2010 Revision." *Waste Management Research* 27, 800–812 (2012); United Nations. "World Population Prospects: The 2015 Revision." *United Nations Economic and Social Affairs* 33, 1–66 (2015); United Nations Department of Economic and Social Affairs Population Division. "World Population Prospects 2024: Online Edition." https://population.un.org/wpp/Download/Standard/CSV/; UN Population Division. "World Population Prospects: The 2010 Revision, CD-ROM Edition" (2011).

29. Yang & Huang. "The Impact of Longevity Risk on the Optimal Contribution Rate and Asset Allocation"; De Waegenaere, A., Melenberg, B. & Stevens, R. "Longevity Risk." *Economist (Leiden)* 158, 151–192 (2010).

30. Koutantou, A. & Papachristou, H. "Greece Pulls the Plug on Pensions for the Dead." *Reuters* (2012).

31. "Greece Paid Pensions to the Dead." BBC News. https://www.bbc.co.uk/news/business-11096512.

32. Koutantou & Papachristou. "Greece Pulls the Plug on Pensions for the Dead."

33. Koutantou & Papachristou. "Greece Pulls the Plug on Pensions for the Dead."

34. Newman. "Supercentenarian and Remarkable Age Records."

35. "Italy's Dead Pensioners." *New York Times* (1997).

Chapter 5

1. Young, R. D. "African American Longevity Advantage: Myth or Reality? A Racial Comparison of Supercentenarian Data." PhD thesis (Georgia State University, 2008). doi: https://doi.org/10.57709/1062194.

2. Young. "African American Longevity Advantage."

3. Young. "African American Longevity Advantage."

4. Young. "African American Longevity Advantage."

5. Young. "African American Longevity Advantage."

6. Young. "African American Longevity Advantage."

7. Young. "African American Longevity Advantage."

8. Young. "African American Longevity Advantage."

9. Perls, T. T., K, B., Freemen, M., Alpert, L. & Silver, M. H. "Age Validation in the New England Centenarian Study" in *Validation of Exceptional Longevity* (Odense University Press, Odense, 1999).

10. Schoenhofen, E. A., Wyszynski, D. F., Andersen, S., et al. "Characteristics of 32 Supercentenarians." *The Journal of the American Geriatrics Society* 54, 1237 (2006).

11. Schoenhofen, Wyszynski, Andersen, et al. "Characteristics of 32 Supercentenarians."

12. Schoenhofen, Wyszynski, Andersen, et al. "Characteristics of 32 Supercentenarians."

13. Robine, J.-M., Gampe, J. & Vaupel, J. "IDL, the International Database on Longevity." *Gerontology* 1–30 (2005); Vaupel, J., Robine, J.-M., Cournil, A., Maier, H. & Gampe, J. "The International Database on Longevity" (2020). https://www.supercentenarians.org/Home/Expand_IDL.

14. Robine, Gampe & Vaupel. "IDL, the International Database on Longevity"; Vaupel, Robine, Cournil, Maier & Gampe. "The International Database on Longevity."

15. Vaupel, Robine, Cournil, Maier & Gampe. "The International Database on Longevity."

16. Vaupel, Robine, Cournil, Maier & Gampe. "The International Database on Longevity."

17. Newman, S. J. "Supercentenarian and Remarkable Age Records Exhibit Patterns Indicative of Clerical Errors and Pension Fraud." *bioRxiv* (2020). doi: https://doi.org/10.1101/704080; Noble, S., McLennan, D., Noble, M., et al. *English Indices of Deprivation 2019: Research Report* (Ministry of Housing Communities & Local Government, London, 2019). https://www.gov.uk/government/publications/english-indices-of-deprivation-2019-research-report.

18. Newman. "Supercentenarian and Remarkable Age Records."

19. Newman. "Supercentenarian and Remarkable Age Records."

20. Ichou, M. & Wallace, M. "The Healthy Immigrant Effect: The Role of Educational Selectivity in the Good Health of Migrants." *Demographic Research* 40, 61–94 (2019).

21. Brumberg, H. L., Dozor, D. & Golombek, S. G. "History of the Birth Certificate: From Inception to the Future of Electronic Data." *Journal of Perinatology* 32, 407–411 (2012).

22. Newman. "Supercentenarian and Remarkable Age Records."

23. Newman. "Supercentenarian and Remarkable Age Records."

24. Newman. "Supercentenarian and Remarkable Age Records."

25. Newman, S. J. & Furbank, R. T. "A Multiple Species, Continent-Wide, Million-Phenotype Agronomic Plant Dataset." *Scientific Data* 8 (2021).

26. Newman. "Supercentenarian and Remarkable Age Records."

27. "Are Supercentenarians Mostly Superfrauds?" Statistical Modeling, Causal Inference, and Social Science. https://statmodeling.stat.columbia.edu/2019/08/06/are-supercentenarians-mostly-superfrauds/.

28. Newman, S. J. "Predicting the Evolution of Life History Using Inclusive Fitness." PhD thesis (Australian National University, 2015).

29. Statistical Modeling, Causal Inference, and Social Science. "Are Supercentenarians Mostly Superfrauds?"; Newman, S. J. "Supercentenarians and the Oldest-Old Are Concentrated into Regions with no Birth Certificates and Short Lifespans." Working paper. https://www.researchgate.net/publication/334497888_Supercentenarians_and_the_oldest-old_are_concentrated_into_regions_with_no_birth_certificates_and_short_lifespans/comments.

30. Statistical Modeling, Causal Inference, and Social Science. "Are Supercentenarians Mostly Superfrauds?"; Newman. "Supercentenarians and the Oldest-Old."

31. Newman. "Supercentenarians and the Oldest-Old."

32. Von Caemmerer, S. & Furbank, R. T. "The C4 Pathway: An Efficient CO2 Pump." *Photosynthetic Research* 77, 191–207 (2003).

33. Newman, S. J. "Ending Government Support for Pro-Alcohol Research." *The Lancet* 393, 1200 (2019); "FOI Docs Released: CSIRO Shapes the Future, but at What Cost?" Drink Tank (2019). http://drinktank.org.au/2019/06/foi-docs-released-csiro-shapes-the-future-but-at-what-cost/; "Australia: Whistleblower Alleges CSIRO Pro-Alcohol Research." Movendi International (2019). https://movendi.ngo/news/2019/04/24/australia-whistleblower-alleges-csiro-pro-alcohol-research/; Davey, M. "Former CSIRO Researcher Accuses Science Agency of Pro-Alcohol Research." *The Guardian* (2019).

34. "Pétition en faveur de l'exhumation des corps de Jeanne et Yvonne Calment—Switzerland." Change.org. https://www.change.org/p/m-emmanuel-macron-p%C3%A9tition-en-faveur-de-l-exhumation-des-corps-de-jeanne-et-yvonne-calment-300e3b0c-4ab7-4bdf-b0bc-df40fc9c6487.

35. Zak, N. & Gibbs, P. "A Bayesian Assessment of the Longevity of Jeanne Calment." *Rejuvenation Research* 23, 3–16 (2020); Zak, N. "Evidence That Jeanne Calment Died in 1934—Not 1997." *Rejuvenation Research* 22, 3–12 (2019).

36. Hoad, P. "'People Are Caught Up in Magical Thinking': Was the Oldest Woman in the World a Fraud?" *The Guardian*. https://www.theguardian.com/science/2019/nov/30/oldest-woman-in-the-world-magical-thinking.

37. Rosenberg, E. "World's Oldest Person: Guinness, 122-Year-Old Jeanne Calment and a Russian Conspiracy Theory." *Washington Post* (2019). https://www.washingtonpost.com/world/2019/01/12/how-madame-calment-worlds-oldest-person-became-fuel-russian-conspiracy-theory/.

38. "Jeanne Calment, une fake news pour l'éternité?" *Le Parisien* (2019). https://www.leparisien.fr/podcasts/code-source/jeanne-calment-une-fake-news-pour-l-eternite-06-05-2019-8066724.php.

39. Hoad. "'People Are Caught Up in Magical Thinking.'"

40. Rosenberg. "World's Oldest Person: Guinness."

41. Rosenberg. "World's Oldest Person: Guinness."

42. Rosenberg. "World's Oldest Person: Guinness."

43. Medvedev, Z. A. "Caucasus and Altay Longevity: A Biological or Social Problem?" *Gerontologist* **14**, 381–387 (1974).

44. Newman, S. J. "Errors as a Primary Cause of Late-Life Mortality Deceleration and Plateaus." *PLoS Biology* **16**, e2006776 (2018); Newman, S. J. "Plane Inclinations: A Critique of Hypothesis and Model Choice in Barbi et al." *PLoS Biology* 16, e3000048 (2018); Newman. *Supercentenarian and Remarkable Age Records*; Newman. "Supercentenarian and Remarkable Age Records."

45. Zak & Gibbs. "A Bayesian Assessment of the Longevity of Jeanne Calment"; Robin-Champigneul, F. "Jeanne Calment's Unique 122-Year Life Span: Facts and

Factors; Longevity History in Her Genealogical Tree." *Rejuvenation Research* 23, 19–47 (2020).

46. Maier, H., Gampe, J., Jeune, B., Robine, J.-M. & Vaupel, J. W. "Supercentenarians." *Demographic Research Monographs* 7 (2010); Jeune, B. & Vaupel, J. W. *Validation of Exceptional Longevity* (Odense University Press, Odense, 1999).

47. Robine, J. M., Allard, M., Herrmann, F. R. & Jeune, B. "The Real Facts Supporting Jeanne Calment as the Oldest Ever Human." *Journals of Gerontology: Series A* 74, S13–S20 (2019).

48. Maier, Gampe, Jeune, Robine & Vaupel. "Supercentenarians."

49. "Was Jeanne Calment the Oldest Person Who Ever Lived—or a Fraud?" *New Yorker* (2020). https://www.newyorker.com/magazine/2020/02/17/was-jeanne-calment-the-oldest-person-who-ever-lived-or-a-fraud.

50. Zak & Gibbs. "A Bayesian Assessment of the Longevity of Jeanne Calment"; Zak. "Evidence That Jeanne Calment Died in 1934—Not 1997."

51. Robine, Allard, Herrmann & Jeune. "The Real Facts Supporting Jeanne Calment as the Oldest Ever Human."

52. Robine, Allard, Herrmann & Jeune. "The Real Facts Supporting Jeanne Calment as the Oldest Ever Human."

53. Davis, N. Z. *The Return of Martin Guerre* (Harvard University Press, Cambridge, MA, 1983).

54. Mannix, L. "Dr Saul Newman Has a Big Idea. It Might Change How We Think About Ageing." *Syndney Morning Herald* (2021). https://www.smh.com.au/national/dr-saul-newman-has-a-big-idea-it-might-change-how-we-think-about-ageing-20210316-p57b32.html.

Chapter 6

1. Buettner, D. "Power 9 Reverse Engineering Longevity." Blue Zones (2016). https://www.bluezones.com/2016/11/power-9/#.

2. Poulain, M., Pes, G. M., Grasland, C., Carru, C., Ferrucci, L., Baggio, G., Franceschi, C. & Deiana, L. "Identification of a Geographic Area Characterized by Extreme Longevity in the Sardinia island: The AKEA Study." *Experimental Gerontology* (2004). https://doi.org/10.1016/j.exger.2004.06.016.

3. Poulain, Pes, Grasland, et al. "Identification of a Geographic Area Characterized by Extreme Longevity."

4. Caselli, G., Pozzi, L., Vaupel, J. W., et al. "Family Clustering in Sardinian Longevity: A Genealogical Approach." *Experimental Gerontology* 41, 727–736 (2006).

5. "Chasing Life." *The Rake*. https://web.archive.org/web/20070928012025/http://www.rakemag.com/stories/section_detail.aspx?itemID=12244&catID=146&SelectCatID=146.

6. "About." Dan Buettner. https://danbuettner.com/about/.

7. Sutton, R. "The Professor of Biking," *Washington Post*, September 20, 1998, https://www.washingtonpost.com/archive/lifestyle/1998/09/21/the-professor-of-biking/7a872e5e-7bde-49f0-a089-3b91e2a15821/.

8. The Rake. "Chasing Life."

9. Buettner, D. *The Blue Zone: Lessons for Living Longer from the People Who've Lived the Longest* (National Geographic, 2008), 304.

10. Buettner, D. "The Secrets of Living Longer." *National Geographic Magazine* 5, 2 (2005).

11. Spector, D. "This Adventurer Discovered The Secrets To Long Life—And It Could Save Iowa $16 Billion By 2016." *Business Insider*, August 8, 2012. https://www.businessinsider.com/dan-buettner-blue-zone-iowa-2012-8.

12. Buettner. "The Secrets of Living Longer."

13. Buettner. "The Secrets of Living Longer."

14. Buettner. "The Secrets of Living Longer."

15. Jeter, C. *Living to 100: Secrets of the Blue Zones* (Netflix, 2023); Buettner, D. & Skemp, S. "Blue Zones: Lessons from the World's Longest Lived." *American Journal of Lifestyle Medicine* 10, 5 (2016). https://doi.org/10.1177/1559827616637066; Poulain, M., Herm, A. & Pes, G. "The Blue Zones: Areas of Exceptional Longevity Around the World." *Vienna Yearbook of Population Research* (2013). https://doi.org/10.1553/populationyearbook2013s87; Buettner, D. *The Blue Zones: Lessons for Living Longer from Those Who've Lived the Longest* (National Geographic Society, Washington, DC, 2010).

16. "De Die Natale." Censorinus. https://elfinspell.com/ClassicalTexts/Maude/Censorinus/DeDieNatale-Part1.html; Thoms, W. J. *Human Longevity, Its Facts and Its Fictions* (John Murray, London, 1873).

17. Leaf, A. & Launois, K. *Youth in Old Age* (McGraw-Hill Companies, Columbus, OH, 1975).

18. Lubalin, P. "To Russia with Yoghurt." *Broadcasting Magazine* (1977), 11.

19. Lubalin. "To Russia with Yoghurt."

20. Speckter, M. "Lerik Journal; Yogurt? Caucasus Centenarians 'Never Eat It.'" *New York Times* (1998). https://www.nytimes.com/1998/03/14/world/lerik-journal-yogurt-caucasus-centenarians-never-eat-it.html.

21. Leaf & Launois. *Youth in Old Age*.

22. Mazess, R. B. & Forman, S. H. "Longevity and Age Exaggeration in Vilcabamba, Ecuador." *Journal of Gerontology* 34, 94–98 (1979).

23. Mazess & Forman. "Longevity and Age Exaggeration in Vilcabamba, Ecuador."

24. Mazess & Forman. "Longevity and Age Exaggeration in Vilcabamba, Ecuador."

25. Mazess, R. B. & Mathisen, R. W. "Lack of Unusual Longevity in Vilcabamba, Ecuador." Human Biology 54, 517–524 (1982). https://www.jstor.org/stable/41463395.

26. Mazess & Forman. "Longevity and Age Exaggeration in Vilcabamba, Ecuador."

27. Hilton, J. *Lost Horizon* (Macmillan & Co., London, 1933).

28. "The Chamber of Commerce and Industry of the Republic of Abkhazia." TPPRA. http://www.tppra.org/en/news/longevity-in-abkhazia-6379.html.

29. Buettner. *The Blue Zone*, ch. 5.

30. "'Blue Zones' Researcher Defends Contested Concept." Colombia One (2025). https://colombiaone.com/2025/01/17/blue-zones-researcher-defends-concept/.

31. Jeter. *Living to 100*; Buettner & Skemp. "Blue Zones"; Buettner. *The Blue Zones*.

32. Jeter. *Living to 100*.

33. "CensusHub." European Statistical System. https://ec.europa.eu/CensusHub/downloadResult.

34. Arias, E., Escobedo, L. A., Kennedy, J., Fu, C. & Cisewski, J. A. "US Small-Area Life Expectancy Estimates Project: Methodology and Results Summary." Centers for Disease Control and Prevention (2018). https://stacks.cdc.gov/view/cdc/58853.

35. "About Family Register Office Work to Affect Location Unknown Elderly People." Japanese Ministry of Justice (2010). http://www.moj.go.jp/MINJI/minji04_00008.html.

36. Poulain, M. & Herm, A. "Exceptional Longevity in Okinawa: Demographic Trends Since 1975." *Journal of Internal Medicine* 295, 387–399 (2024).

37. Glei, D. A., Barbieri, M. & Santamaría-Ulloa, C. "Costa Rican Mortality 1950–2013: An Evaluation of Data Quality and Trends Compared with Other Countries." *Demographic Research* 40, 835 (2019); Bixby, L., Brenes, G. & Collado, A. "Tablas de vida para cálculo actuarial de rentas vitalicias y retiro programado Costa Rica circa 2000." *Población y Salud en Mesoamérica* 173–205 (2004).

38. Glei, Barbieri & Santamaría-Ulloa. "Costa Rican Mortality 1950–2013."

39. Glei, Barbieri & Santamaría-Ulloa. "Costa Rican Mortality 1950–2013."

40. Rosero-Bixby, L. "The Vanishing Advantage of Longevity in Nicoya, Costa Rica: A Cohort Shift." *Demographic Research* 49 (2023).

41. Arias, Escobedo, Kennedy, Fu & Cisewski. "US Small-Area Life Expectancy Estimates Project"; Bixby, Brenes & Collado. "Tablas de vida para cálculo actuarial de rentas vitalicias y retiro programado"; "Life Expectancy by Age, Sex and NUTS 2 Region." Eurostat (2023). https://ec.europa.eu/eurostat/databrowser/product/page/demo_r_mlifexp; "Trends in Vital Statistics by Prefecture in Japan, 1899–1998." Ministry of Health, Labor and Welfare of Japan. https://www.mhlw.go.jp/english/database/db-hw/vs_8/index.html.

42. Dewitt, D. "The $3 Million Brevard Blue Zones Project: A Mix of Real Benefits and Marketing Claims." Brevard NewsBeat (2023). https://brevardnewsbeat.substack.com/p/the-3-million-brevard-blue-zones.

43. Jeter. *Living to 100*; Siu, E. "Singapore Has 'Engineered' a Society That Lives to 100—Here's How." CNBC (2024). https://www.cnbc.com/2024/03/27/singapore-has-engineered-a-society-that-lives-to-100-heres-how.html.

44. "Registration of Births and Deaths Act." Singapore Statutes Online (2021). https://sso.agc.gov.sg/Act-Rev/RBDA1937/Published/20211231?DocDate=19870330&ViewType=Within.

45. Descalsota, M. & Tan, K. W. "Singapore Is the Newest Blue Zone, but Some Locals Disagree." Business Insider (2023). https://www.businessinsider.com/singapore-new-blue-zone-netflix-locals-disagree-life-sucks-2023-9.

46. Finkelstein, P. "'Blue Zones' Have Captivated Health and Longevity Experts. But Are They Real or Statistical Grift?" *Salon*, October 1, 2023. https://www.salon.com/2023/10/01/blue-zones-have-captivated-health-and-longevity-experts-but-are-they-real-or-statistical-grift/.

47. "Speech by Mr Ong Ye Kung, Minister for Health." Singapore Ministry of Health (2023). https://www.moh.gov.sg/news-highlights/details/speech-by-mr-ong-ye-kung-minister-for-health-at-the-3rd-centre-for-research-on-successful-ageing-symposium-on-thursday-12-october-2023-9.45am.

48. Ministry of Health, Labor and Welfare of Japan. "Trends in Vital Statistics by Prefecture in Japan, 1899–1998"; "The Survey on Japanese Value Orientations, 1973–2008." Public Opinion Research Division, NHK Broadcasting Culture Research Institute (2023). doi:10.34500/SSJDA.0814; "e-Stat: Portal Site of Official Statistics of Japan." Statistics Japan (2020). https://www.e-stat.go.jp/en; Ikeda, N., Nakaya, T., Bennett, J., Ezzati, M. & Nishi, N. "Trends and Disparities in Adult Body Mass Index Across the 47 Prefectures of Japan, 1975–2018: A Bayesian Spatiotemporal Analysis of National Household Surveys." *Frontiers in Public Health* 10, 830578 (2022); "Family Income and Expenditure Survey." Statistics Bureau of Japan (2016). https://www.stat.go.jp/english/data/kakei/index.html; "Survey on Time Use and Leisure

Activities." Statistics Bureau of Japan (2016). https://www.stat.go.jp/english/data/shakai/2016/h28kekka.html.

49. Buettner. "Power 9 Reverse Engineering Longevity."

50. Ikeda, Nakaya, Bennett, Ezzati & Nishi. "Trends and Disparities in Adult Body Mass Index Across the 47 Prefectures of Japan."

51. Ikeda, Nakaya, Bennett, Ezzati & Nishi. "Trends and Disparities in Adult Body Mass Index."

52. Ikeda, Nakaya, Bennett, Ezzati & Nishi. "Trends and Disparities in Adult Body Mass Index."

53. Statistics Bureau of Japan. "Survey on Time Use and Leisure Activities."

54. Statistics Bureau of Japan. "Survey on Time Use and Leisure Activities."

55. Buettner. "Power 9 Reverse Engineering Longevity."

56. Buettner. "Power 9 Reverse Engineering Longevity."

57. Jeter. *Living to 100*; Buettner & Skemp. "Blue Zones"; Buettner. *The Blue Zones*.

58. Statistics Bureau of Japan. "Family Income and Expenditure Survey"; Ministry of Health Labor and Welfare. *National Health and Nutrition Survey Report* (Ministry of Health Labor and Welfare, Tokyo, 2016). https://www.nibiohn.go.jp/eiken/kenkounippon21/en/eiyouchousa/; "Consumption Expenditure of Seafood." Statistics Japan. https://stats-japan.com/t/kiji/14620.

59. OECD-FAO. *Agricultural Outlook 2015–2024* (OECD Publishing, Paris, 2015).

60. Motomura, C. *Global Agricultural Information Network Regional Report—Okinawa* (USDA Foreign Agricultural Service, Washington, DC, 2014).

61. Statistics Japan. "Consumption Expenditure of Seafood."

62. Statistics Bureau of Japan. "Family Income and Expenditure Survey"; Ministry of Health Labor and Welfare. *National Health and Nutrition Survey Report*.

63. Statistics Bureau of Japan. "Family Income and Expenditure Survey"; Ministry of Health Labor and Welfare. *National Health and Nutrition Survey Report*.

64. Statistics Bureau of Japan. "Family Income and Expenditure Survey"; Ministry of Health Labor and Welfare. *National Health and Nutrition Survey Report*; "Consumption Expenditure of Salad." Statistics Japan. https://stats-japan.com/t/kiji/24476.

65. Buettner. "Power 9 Reverse Engineering Longevity."

66. Public Opinion Research Division, NHK Broadcasting Culture Research Institute. "The Survey on Japanese Value Orientations, 1973–2008."

67. "Statistics on Religion Survey." Ministry of Education, Culture, Sports, Science and Technology (2016). https://www.mext.go.jp/b_menu/toukei/chousa07/shuukyou/1262852.htm.

68. Buettner. "Power 9 Reverse Engineering Longevity."

69. Ministry of Health, Labor and Welfare of Japan. "Trends in Vital Statistics by Prefecture in Japan, 1899–1998."

70. "System of Social and Demographic Statistics (SDSS)." Ministry of Internal Affairs and Communications Statistics Bureau (2023). https://www.e-stat.go.jp/en/regional-statistics/ssdsview.

71. Buettner. "Power 9 Reverse Engineering Longevity."

72. Ministry of Health, Labor and Welfare of Japan. "Trends in Vital Statistics by Prefecture in Japan, 1899–1998"; Ministry of Internal Affairs and Communications Statistics Bureau. "System of Social and Demographic Statistics."

73. Ministry of Internal Affairs and Communications Statistics Bureau. "System of Social and Demographic Statistics."

74. Ministry of Health, Labor and Welfare of Japan. "Trends in Vital Statistics by Prefecture in Japan, 1899–1998."

75. Ministry of Internal Affairs and Communications Statistics Bureau. "System of Social and Demographic Statistics."

76. Buettner. "Power 9 Reverse Engineering Longevity."

77. "Statistics on Imposition of Liquor Tax 2016." National Tax Agency (2016). https://www.nta.go.jp/taxes/sake/tokei/mokuji.htm.

78. World Health Organization. *Global Status Report on Alcohol and Health* (World Health Organization, Geneva, 2018); Global Burden of Disease. "Global, Regional, and National Age-Sex Specific Mortality for 264 Causes of Death, 1980–2016: A Systematic Analysis for the Global Burden of Disease Study 2016." *The Lancet* 390, 1151–1210 (2017).

79. Buettner. "Power 9 Reverse Engineering Longevity."

80. Buettner. "Power 9 Reverse Engineering Longevity."

81. Ministry of Health Labor and Welfare. *National Health and Nutrition Survey Report*; Ministry of Internal Affairs and Communications Statistics Bureau. "System of Social and Demographic Statistics"; "Tobacco." World Health Organization (2025). https://www.who.int/news-room/fact-sheets/detail/tobacco.

82. Ikeda, Nakaya, Bennett, Ezzati & Nishi. "Trends and Disparities in Adult Body Mass Index."

83. “Patients with Major Depressive Disorder (Aged 65 or Above).” Statistics Japan. https://stats-japan.com/t/kiji/14955.

84. Poulain, M. “Exceptional Longevity in Okinawa: A Plea for In-Depth Validation.” *Demographic Research* (2011). doi:10.4054/DemRes.2011.25.7.

85. Trabidou, G. & Florou, H. “Estimation of Dose Rates to Humans Exposed to Elevated Natural Radioactivity Through Different Pathways in the Island of Ikaria, Greece.” *Radiation Protection Dosimetry* 142, 378–384 (2010); Trabidou, G., Florou, H., Angelopoulos, A. & Sakelliou, L. “Environmental Study of the Radioactivity of the Spas in the Island of Ikaria.” *Radiation Protection Dosimetry* 63, 63–67 (1996).

86. Trabidou, Florou, Angelopoulos & Sakelliou. “Environmental Study of the Radioactivity.”

87. Trabidou & Florou. “Estimation of Dose Rates to Humans Exposed to Elevated Natural Radioactivity.”

88. Trabidou, Florou, Angelopoulos & Sakelliou. “Environmental Study of the Radioactivity.”

89. Papalas, A. J. *Ancient Icaria* (Bolchazy-Carducci Publishers, Wauconda, 1992); Papalas, A. J. *Rebels and Radicals: Icaria 1600–2000* (Bolchazy Carducci Publishers, Wauconda, 2005).

90. Koutantou, A. & Papachristou, H. “Greece Pulls the Plug on Pensions for the Dead.” *Reuters* (2012); “Greece Paid Pensions to the Dead.” BBC News. https://www.bbc.co.uk/news/business-11096512; Papalas. *Rebels and Radicals*.

91. Rosero-Bixby. “The Vanishing Advantage of Longevity in Nicoya, Costa Rica.”

92. Descalsota & Tan. “Singapore Is the Newest Blue Zone, but Some Locals Disagree.”

93. Ikeda, Nakaya, Bennett, Ezzati & Nishi. “Trends and Disparities in Adult Body Mass Index.”

94. Ikeda, Nakaya, Bennett, Ezzati & Nishi. “Trends and Disparities in Adult Body Mass Index.”

95. “Do People in ‘Blue Zones’ Really Live Longer?” *New York Times* (2024). https://www.nytimes.com/2024/10/24/well/live/blue-zones-longevity-aging.html.

96. Amigo, I. “Do Blue Zones, Supposed Havens Longevity, Rest on Shaky Science.” *Science* 386, 6724 (2024).

97. “Lessons in Longevity from Japan’s 100-Year-Old Okinawans.” World Economic Forum (2021). https://www.weforum.org/stories/2021/09/japan-okinawa-secret-to-longevity-good-health/; Willcox, B. J., Willcox, D. C., Suzuki, M. *The Okinawa Diet Plan: Get Leaner, Live Longer, and Never Feel Hungry* (Harmony, New York, 2005).

98. Trademark Assignment Cover Sheet 900995124. Assignment ID TMI1531362. https://assignmentcenter.uspto.gov/ipas/search/api/v2/public/download/trademark/9132/0805.

99. Fletcher, H. "Healthways to Sell Division, Brand to ShareCare in Major Corporate Shake-Up." *The Tennessean* (2016). https://eu.tennessean.com/story/money/industries/health-care/2016/07/27/healthways-sells-population-health-division-brand-to-sharecare/87631186/.

100. "James Randi Speaks Pigasus Awards 2009." YouTube. https://www.youtube.com/watch?v=iujAqCpBf_k; Mestel, R. "Dr. Oz, Andrew Wakefield and Others, Um, 'Honored' by James Randi." *Los Angeles Times* (2011). https://www.latimes.com/health/la-xpm-2011-apr-01-la-heb-dr-oz-andrew-wakefield-james-randi-awards-20110401-story.html; "JREF's Pigasus Awards 'Honors' Dubious Peddlers of 'Woo' (VIDEO)." James Randi Educational Foundation (2013). http://archive.randi.org/site/index.php/jref-news/2074-jrefs-pigasus-awards-to-honor-dubious-peddlers-of-woo.html.

101. Cohen, P. "Group of Doctors Calls on Columbia Univ. to Oust Dr. Oz." CBS News (2015). https://www.cbsnews.com/news/group-of-doctors-call-for-dr-oz-to-be-ousted-from-columbia-university/.

102. Margolies, J. "Does This Brooklyn Development Know the Secret to Long Life?" *The New York Times*, July 28, 2023. https://www.nytimes.com/2023/07/28/nyregion/new-york-brooklyn-blue-zones.html.

103. "Cedar Rapids and Marion Dropping Blue Zones, Joining Healthy Hometown Instead." *The Gazette* (2018). https://www.thegazette.com/government-politics/cedar-rapids-and-marion-dropping-blue-zones-joining-healthy-hometown-instead/; Love, S. "The Longevity Hot Spots That Weren't." *The New Republic* (2024). https://www.yahoo.com/lifestyle/longevity-hot-spots-weren-t-110000336.html.

104. Love, S. "The Longevity Hot Spots That Weren't."

105. Dewitt. "The $3 Million Brevard Blue Zones Project."

106. Love. "The Longevity Hot Spots That Weren't."

107. Love. "The Longevity Hot Spots That Weren't."

108. Dewitt. "The $3 Million Brevard Blue Zones Project."

109. Boutron, I., Moher, D., Altman, D. G., Schulz, K. F. & Ravaud, P. "Extending the CONSORT Statement to Randomized Trials of Nonpharmacologic Treatment: Explanation and Elaboration." *Annals of Internal Medicine* 148, 295–309 (2008); "Ethical Considerations" in *Field Trials of Health Interventions: A Toolbox* (National Center for Biotechnology Information, Oxford, 2015). https://www.ncbi.nlm.nih.gov/books/NBK305506/.

110. Jeter. *Living to 100.*

111. Buettner & Skemp. "Blue Zones."

112. Buettner & Skemp. "Blue Zones."

113. Buettner & Skemp. "Blue Zones."

114. Love. "The Longevity Hot Spots That Weren't"; "Blue Zones and Healthly Partner to Focus Healthcare on the Business of Reversing Disease." PR Newswire (2024). https://www.prnewswire.com/news-releases/blue-zones-and-healthly-partner-to-focus-healthcare-on-the-business-of-reversing-disease-302203260.html.

115. Love. "The Longevity Hot Spots That Weren't."

116. Fletcher. "Healthways to Sell Division, Brand to ShareCare in Major Corporate Shake-Up"; PR Newswire. "Blue Zones and Healthly Partner to Focus Healthcare on the Business of Reversing Disease"; "Can Your City Make You Healthy? Blue Zones Project Communities say 'Yes.'" PR Newswire (2016). https://www.prnewswire.com/news-releases/can-your-city-make-you-healthy-blue-zones-project-communities-say-yes-300330086.html. See also Blue Zones. "Measuring the Benefits of Better Health." (2026). https://www.bluezonesproject.com/measurement/.

117. "Blue Zones and American College of Lifestyle Medicine Create Partnership to Accelerate a Lifestyle-First Approach to All Healthcare." PR Newswire (2024). https://www.prnewswire.com/news-releases/blue-zones-and-american-college-of-lifestyle-medicine-create-partnership-to-accelerate-a-lifestyle-first-approach-to-all-healthcare-302144834.html.

118. PR Newswire. "Blue Zones and Healthly Partner to Focus Healthcare on the Business of Reversing Disease."

Chapter 7

1. Maier, H., Gampe, J., Jeune, B., Robine, J.-M. & Vaupel, J. W. "Supercentenarians." *Demographic Research Monographs* 7 (2010); Charbonneau, H. "Pierre Joubert a-t-il vécu 113 ans?" *Mémoires de la Société généalogique canadienne-française* 41, 45–48 (1990).

2. "Russian Researcher Questions Truth Behind France's Oldest Woman." Radio France Internationale (2019). https://www.rfi.fr/en/france/20190102-russian-researchers-question-truth-behind-frances-oldest-woman.

3. Hopper, T. "History's Oldest Woman a Fraud? Russian Researchers Claim 122-Year-Old Jeanne Calment Was Actually a 99-Year-Old Imposter." *National Post.* https://nationalpost.com/news/world/historys-oldest-woman-a-fraud-theory-says-122-year-old-jeanne-calment-was-actually-a-99-year-old-imposter.

4. Vallin, J. "Why Are Supercentenarians So Frequently Found in French Overseas Departments? The Cases of Guadeloupe and Martinique." *Genus* 76, 1–17 (2020).

5. Vallin. "Why Are Supercentenarians So Frequently Found in French Overseas Departments?"

6. Vallin. "Why Are Supercentenarians So Frequently Found in French Overseas Departments?"

7. Markides, K. S. & Machalek, R. "Selective Survival, Aging and Society." *Archives of Gerontology and Geriatrics* **3**, 207–222 (1984).

8. Karima, L. "North African Diaspora in France and Colonial Legacies" in *Diasporas: Movements and Cultures* (Critical, Cultural and Communication Press, Nottingham, 2007).

9. Karima, L. "North African Diaspora in France and Colonial Legacies"; Varus, R. "The Role of France and the French in North Africa." *Annuals of the American Academy of Political and Social Science* 306, 17–25 (1956).

10. Jensen, E., Roberts, A. & Rogers, L. "Age Heaping in the 2020 Census Demographic and Housing Characteristics File (DHC)." US Census Bureau (2023). https://www.census.gov/newsroom/blogs/random-samplings/2023/05/age-heaping-2020-census-dhc.html.

11. Jensen, Roberts & Rogers. "Age Heaping in the 2020 Census Demographic and Housing Characteristics File."

12. Newman, S. J. "Errors as a Primary Cause of Late-Life Mortality Deceleration and Plateaus." *PLoS Biology* 16, e2006776 (2018).

13. Shapiro, S. & Schachter, J. "Birth Registration Completeness, United States, 1950." *Public Health Reports* 67, 513–524 (1952).

14. Shapiro & Schachter. "Birth Registration Completeness, United States, 1950."

15. Brumberg, H. L., Dozor, D. & Golombek, S. G. "History of the Birth Certificate: From Inception to the Future of Electronic Data." *Journal of Perinatology* 32, 407–411 (2012).

16. Brumberg, Dozor & Golombek. "History of the Birth Certificate: From Inception to the Future of Electronic Data"; Shapiro & Schachter. "Birth Registration Completeness, United States, 1950."

17. Hambright, T. Z. "Comparison of Information on Death Certificates and Matching 1960 Census Records: Age, Marital Status, Race, Nativity and Country of Origin." *Demography* 6, 413–423 (1969).

18. Hambright. "Comparison of Information on Death Certificates and Matching 1960 Census Records."

19. Hambright. "Comparison of Information on Death Certificates and Matching 1960 Census Records."

20. Hambright. "Comparison of Information on Death Certificates and Matching 1960 Census Records."

21. US Department of Health and Human Services, Office of Inspector General. "Birth Certificate Fraud." OEI-07-99-00570 (2000). http://www.hhs.gov/oig/oei/.

22. US Department of Health and Human Services, Office of Inspector General. "Birth Certificate Fraud."

23. US Department of Health and Human Services, Office of Inspector General. "Birth Certificate Fraud."

24. US Department of Health and Human Services, Office of Inspector General. "Birth Certificate Fraud."

25. US Department of Health and Human Services, Office of Inspector General. "Birth Certificate Fraud."

26. Newman. "Errors as a Primary Cause of Late-Life Mortality Deceleration and Plateaus"; Newman, S. J. "Supercentenarian and Remarkable Age Records Exhibit Patterns Indicative of Clerical Errors and Pension Fraud." *bioRxiv* (2020). doi: https://doi.org/10.1101/704080.

27. Legislative Assembly of Puerto Rico. "Law 191 of 2009: Law Prohibiting Public and Private Entities from Retaining, Storing, or Holding Certified Copies of Birth Certificates" (2009). https://ilw.com/immigrationdaily/news/2010,0507-PRbirth.pdf; Rosenwaike, I. & Preston, S. H. "Age Overstatement and Puerto Rican Longevity." *Human Biology* 56, 502–525 (1984).

28. Legislative Assembly of Puerto Rico. "Law 191 of 2009."

29. Taj El Din, S. A. "Age Estimation Committee in Qatar." International Institute for Vital Registration and Statistics (1980). https://unstats.un.org/unsd/demographic-social/crvs/documents/IIVRS_papers/IIVRS_paper12.pdf.

30. Cappa, C., Gregson, K., Wardlaw, T. & Bissell, S. "Birth Registration: A Child's Passport to Protection." *Lancet Global Health* 2 (2014). https://doi.org/10.1016/S2214-109X(13)70180-3; United Nations. "Coverage of Birth and Death Registration." *United Nations Demographic Yearbook 2015: Quality of Vital Statistics Obtained from Civil Registration* (2017). https://unstats.un.org/unsd/demographic/CRVS/CR_coverage.htm; UNICEF. *Birth Registration for Every Child by 2030: Are We on Track?* (UNICEF, New York, 2019).

31. United Nations. "Coverage of Birth and Death Registration."

32. Cappa, Gregson, Wardlaw & Bissell. "Birth Registration: A Child's Passport to Protection"; United Nations. "Coverage of Birth and Death Registration."

33. United Nations. "Coverage of Birth and Death Registration"; Phillips, D. E., Lozano, R., Naghavi, M., et al. "A Composite Metric for Assessing Data on Mortality and Causes of Death: The Vital Statistics Performance Index." *Population Health Metrics* 12, 14 (2014).

34. Saikia, N., Kumar, K. & Das, B. "Death Registration Coverage 2019–2021, India." *Bulletin of the World Health Organisation* 101, 102 (2022).

35. Lenzer, J. "Study of London Taxi Drivers Wins Ig Nobel Prize." *BMJ: British Medical Journal* 327, 831 (2003).

36. Newman. "Errors as a Primary Cause of Late-Life Mortality Deceleration and Plateaus."

37. Newman, S. J. "The Global Pattern of Centenarians Highlights Deep Problems in Demography." *medRxiv* (2024). doi:10.1101/2024.09.06.24313170.

38. Newman. "The Global Pattern of Centenarians Highlights Deep Problems in Demography."

39. Newman. "The Global Pattern of Centenarians Highlights Deep Problems in Demography."

40. Newman. "The Global Pattern of Centenarians Highlights Deep Problems in Demography."

41. Newman. "The Global Pattern of Centenarians Highlights Deep Problems in Demography."

42. Jensen, Roberts & Rogers. "Age Heaping in the 2020 Census Demographic and Housing Characteristics File."

43. Newman, S. J. "Early-Life Physical Performance Predicts the Ageing and Death of Elite Athletes. *Science Advances* 9 (2023); Newman, S. J., Schulte, K., Morellini, M. M., Rahal, C. & Leasure, D. R. "Offshoring Emissions Through Used Vehicle Exports." *Nature Climate Change* 14, 238–241 (2024). Victor, B., Nibali, A., Newman, S. J., et al. "High-Throughput Plot-Level Quantitative Phenotyping Using Convolutional Neural Networks on Very High-Resolution Satellite Images." *Remote Sensing* 16, 282 (2024); Furbank, R. T., Silva-Perez, V., Evans, J. R., et al. "Wheat Physiology Predictor: Predicting Physiological Traits in Wheat from Hyperspectral Reflectance Measurements Using Deep Learning." *Plant Methods* 17, 1–15 (2021).

44. Newman, Schulte, Morellini, Rahal & Leasure. "Offshoring Emissions Through Used Vehicle Exports."

Chapter 8

1. Kaplan, H., Thompson, R. C. & Trumble, B. C., et al. "Coronary Atherosclerosis in Indigenous South American Tsimane: A Cross-Sectional Cohort Study." *The Lancet*

389, 1730–1739 (2017); Kaplan, H., Hooper, P. L., Gatz, M., et al. "Brain Volume, Energy Balance, and Cardiovascular Health in Two Nonindustrial South American Populations." *Proceedings of the National Academy of Sciences* 120 (2023).

2. Mazess, R. B. & Forman, S. H. "Longevity and Age Exaggeration in Vilcabamba, Ecuador." *Journal of Gerontology* 34, 94–98 (1979); Mazess, R. B. & Mathisen, R. W. "Lack of Unusual Longevity in Vilcabamba, Ecuador." *Human Biology* 54, 517–524 (1982). https://www.jstor.org/stable/41463395.

3. "The Community Where People Age More Slowly Than the Rest of the World." BBC News. https://www.bbc.co.uk/news/articles/ceq55l2gdxxo.

4. Gurven, M., Stieglitz, J., Trumble, B., et al. "The Tsimane Health and Life History Project: Integrating Anthropology and Biomedicine." *Evolutionary Anthropology* 26, 54 (2017); Central Intelligence Agency. *CIA World Factbook: Life Expectancy at Birth* (CIA, Washington, DC, 2022).

5. BBC News. "The Community Where People Age More Slowly Than the Rest of the World"; Leonard, W. R., Reyes-Garcia, V., Tanner, S., et al. "The Tsimane' Amazonian Panel Study (TAPS): Nine Years (2002–2010) of Annual Data Available to the Public." *Economics and Human Biology* 19, 51–61 (2015).

6. "Tsimane Amazonian Panel Study: Data Sets." Brandeis University. https://heller.brandeis.edu/sustainable-international-development/tsimane/data.html.

7. Brandeis University. "Tsimane Amazonian Panel Study: Data Sets."

8. Gurven, Stieglitz, Trumble, et al. "The Tsimane Health and Life History Project."

9. Kaplan, Thompson & Trumble. "Coronary Atherosclerosis in Indigenous South American Tsimane."

10. BBC News. "The Community Where People Age More Slowly Than the Rest of the World."

11. Tolu, F., Palermo, M., Pina Dore, M., et al. "Association of Endemic Goitre and Exceptional Longevity in Sardinia: Evidence from an Ecological Study." *European Journal of Ageing* 16, 405–414 (2019).

Chapter 9

1. Newman, S. J. & Easteal, S. "Global Patterns of Human Ageing." *bioRxiv* (2019). doi: https://doi.org/10.1101/124792.

2. Newman & Easteal. "Global Patterns of Human Ageing."

3. Kubin, M., Wagner, G. & Fugl-Meyer, A. R. "Epidemiology of erectile dysfunction." *International Journal of Impotence Research* 15, 1 63–71 (2003).

4. Di Francesco, A., Deighan, A. G., Litichevskiy, L., et al. "Dietary restriction impacts health and lifespan of genetically diverse mice." *Nature* 634, 684–692 (2024).

5. Di Francesco, Deighan, Litichevskiy, et al. "Dietary restriction impacts health and lifespan of genetically diverse mice."

6. Di Francesco, Deighan, Litichevskiy, et al. "Dietary restriction impacts health and lifespan of genetically diverse mice."

7. Harrison, D. E., Strong, R., Sharp, Z. D., Nelson, J. F., Astle, C. M., Flurkey, K., Nadon, N. L., Wilkinson, J. E., Frenkel, K., Carter, C. S., Pahor, M., Javors, M. A., Fernandez, E., & Miller, R. A. (2009). "Rapamycin fed late in life extends lifespan in genetically heterogeneous mice." *Nature* 460, 392. doi:10.1038/NATURE08221.

8. Miller, R. A., Strong, R., Korstanje, R., Harrison, D. E. & Rosenthal, N. A. "ITP: Interventions Testing Program: Effects of Various Treatments on Lifespan and Related Phenotypes in Genetically Heterogeneous Mice (UM-HET3) (2004–2024)." Mouse Phenome Database at the Jackson Library. https://phenome.jax.org/projects/ITP1.

9. Miller, R. A., Harrison, D. E., Astle, C. M., et al. "An Aging Interventions Testing Program: Study Design and Interim Report." *Aging Cell* 6, 565–575 (2007).

10. Strong, R., Miller, R. A., Astle, C. M., et al. "Nordihydroguaiaretic Acid and Aspirin Increase Lifespan of Genetically Heterogeneous Male Mice." *Aging Cell* 7, 641 (2008); Miller, Strong, Korstanje, Harrison & Rosenthal. "ITP: Interventions Testing Program."

11. Strong, Miller, Astle, et al. "Nordihydroguaiaretic Acid and Aspirin Increase Lifespan of Genetically Heterogeneous Male Mice."

12. De Magalhães, J. P., Thompson, L., de Lima, I., et al. "A Reassessment of Genes Modulating Aging in Mice Using Demographic Measurements of the Rate of Aging." *Genetics* 208, 1617–1630 (2018).

Chapter 10

1. Couzin, J. "Aging Research's Family Feud." *Science* 303, 1276–1279 (2004).

2. "Is Elysium Health's Basis the Fountain of Youth?" Science of Us (2016). https://www.thecut.com/2016/08/is-elysium-healths-basis-the-fountain-of-youth.html.

3. Couzin. "Aging Research's Family Feud."

4. Science of Us. "Is Elysium Health's Basis the Fountain of Youth?"

5. Husten, L. "Resveratrol and Fraud." *Forbes* (2012). https://www.forbes.com/sites/larryhusten/2012/01/16/resveratrol-and-fraud/.

6. Couzin. "Aging Research's Family Feud"; "A Clinical Study to Assess the Safety and Activity of SRT501 Alone or in Combination with Bortezomib in Patients with

Multiple Myeloma." ClinicalTrials.gov (2019). https://clinicaltrials.gov/study/NCT00920556?tab=results.

7. "David Sinclair at TEDMED 2009." YouTube. https://www.youtube.com/watch?v=xCGwhXOP8_Y.

8. "A 'Fountain of Youth' Pill? Sure, If You're a Mouse." KFF Health News. https://kffhealthnews.org/news/a-fountain-of-youth-pill-sure-if-youre-a-mouse/.

9. Walle, T. "Bioavailability of Resveratrol." *Annals of the New York Academy of Sciences* 1215, 9–15 (2011).

10. Pezzuto, J. M. "Resveratrol: Twenty Years of Growth, Development and Controversy." *Biomolecules & Therapeutics (Seoul)* 27, 1 (2018).

11. Brown, V. A., Patel, K. R., Viskaduraki, M., et al. "Repeat Dose Study of the Cancer Chemopreventive Agent Resveratrol in Healthy Volunteers: Safety, Pharmacokinetics, and Effect on the Insulin-Like Growth Factor Axis." *Cancer Research* 70, 9003–9011 (2010).

12. Koushki, M., Amiri-Dashatan, N., Ahmadi, N., Abbaszadeh, H. A. & Rezaei-Tavirani, M. "Resveratrol: A Miraculous Natural Compound for Diseases Treatment." *Food Science & Nutrition* 6, 2473 (2018); Brown, Patel, Viskaduraki, et al. "Repeat Dose Study of the Cancer Chemopreventive Agent Resveratrol in Healthy Volunteers."

13. ClinicalTrials.gov. "A Clinical Study to Assess the Safety and Activity of SRT501."

14. ClinicalTrials.gov. "A Clinical Study to Assess the Safety and Activity of SRT501."

15. Lowe, D. "Resveratrol (SRT501): Development Halted." Science.org (2010). https://www.science.org/content/blog-post/resveratrol-srt501-development-halted.

16. ClinicalTrials.gov. "A Clinical Study to Assess the Safety and Activity of SRT501."

17. Wade, N. "Doubt on Anti-Aging Molecule as Resveratrol Trial Is Halted." *New York Times* (2011). https://www.nytimes.com/2011/01/11/science/11aging.html.

18. Koushki, Amiri-Dashatan, Ahmadi, Abbaszadeh & Rezaei-Tavirani. "Resveratrol: A Miraculous Natural Compound for Diseases Treatment."

19. Wade. "Doubt on Anti-Aging Molecule as Resveratrol Trial Is Halted."

20. Pezzuto. "Resveratrol: Twenty Years of Growth, Development and Controversy."

21. Pezzuto. "Resveratrol: Twenty Years of Growth, Development and Controversy."

22. Berman, A. Y., Motechin, R. A., Wiesenfeld, M. Y. & Holz, M. K. "The Therapeutic Potential of Resveratrol: A Review of Clinical Trials." *NPJ Precision Oncology* **1**, 1–9 (2017).

23. Husten. "Resveratrol and Fraud."

24. Couzin. "Aging Research's Family Feud."

25. Lowe, D. "Resveratrol (SRT501): Buy Now—Why Wait?" Science.org (2010). https://www.science.org/content/blog-post/resveratrol-srt501-buy-now---why-wait.

26. "So How Peripheral Was Dipak Das' Resveratrol Work, Really?" Retraction Watch (2012). https://retractionwatch.com/2012/01/12/so-how-peripheral-was-dipak-das-resveratrol-work-really/.

27. Jaslow, R. "Red Wine Researcher Dr. Dipak K. Das Published Fake Data: UConn." CBS News (2012). https://www.cbsnews.com/news/red-wine-researcher-dr-dipak-k-das-published-fake-data-uconn/.

28. KFF Health News. "A 'Fountain of Youth' Pill? Sure, If You're a Mouse."

29. Weintraub, K. "Rejuvenating the Chance of Motherhood?" *MIT Technology Review* (2016). https://www.technologyreview.com/2016/12/09/155229/rejuvenating-the-chance-of-motherhood/.

30. KFF Health News. "A 'Fountain of Youth' Pill? Sure, If You're a Mouse."

31. KFF Health News. "A 'Fountain of Youth' Pill? Sure, If You're a Mouse."

32. "Live Healthier Longer Course." InsideTracker. https://info.insidetracker.com/live-healthier-longer-course?_gl=1*1sygfxw*_gcl_au*MTU0NDU4NjAxNy4xNzM1NTE2MzE1.

33. KFF Health News. "A 'Fountain of Youth' Pill? Sure, If You're a Mouse."

34. Sharoni, E. "A Milestone for Personalized Nutrition: Our First Published Paper Validates InsideTracker's Effectiveness." InsideTracker (2021). https://blog.insidetracker.com/milestone-for-personalized-nutrition-our-first-paper-published-validates-insidetracker-effectiveness.

35. Westerman, K., Reaver, A., Roy, C., et al. "Longitudinal Analysis of Biomarker Data from a Personalized Nutrition Platform in Healthy Subjects." *Scientific Reports* 8 (2018).

36. Westerman, Reaver, Roy, et al. "Longitudinal Analysis of Biomarker Data."

37. Westerman, Reaver, Roy, et al. "Longitudinal Analysis of Biomarker Data."

38. "CohBar, Inc.: Notice of Delisting or Failure to Satisfy a Continued Listing Rule or Standard; Transfer of Listing (Form 8-K)." CohBar. https://www.cohbar.com/news-media/press-releases/detail/187/cohbar-inc-notice-of-delisting-or-failure-to-satisfy-a.

39. Dobkin, R. "New Supplement Reverses Aging in Dogs, Study Shows." Newsweek (2024). https://www.newsweek.com/new-supplement-reverses-aging-dogs-study-1874899.

40. David A. Sinclair on X. https://x.com/davidasinclair/status/1763267566582084053.

41. Simon, K. E., Russell, K., Mondino, A., et al. "A Randomized, Controlled Clinical Trial Demonstrates Improved Owner-Assessed Cognitive Function in Senior Dogs Receiving a Senolytic and NAD+ Precursor Combination." *Scientific Reports* 14, 1–16 (2024).

42. "A Randomized, Controlled Clinical Trial Demonstrates Improved Cognitive Function in Senior Dogs Supplemented with a Senolytic and NAD+ Precursor Combination." PubPeer. https://pubpeer.com/publications/0463F989D32D0CCE55803A0FF67D87?utm_source=Firefox&utm_medium=BrowserExtension&utm_campaign=Firefox.

43. Simon, Russell, Mondino, et al. "A Randomized, Controlled Clinical Trial Demonstrates Improved Owner-Assessed Cognitive Function in Senior Dogs."

44. Simon, Russell, Mondino, et al. "A Randomized, Controlled Clinical Trial Demonstrates Improved Owner-Assessed Cognitive Function in Senior Dogs."

45. Simon, Russell, Mondino, et al. "A Randomized, Controlled Clinical Trial Demonstrates Improved Owner-Assessed Cognitive Function in Senior Dogs."

46. PubPeer. "A Randomized, Controlled Clinical Trial Demonstrates Improved Cognitive Function in Senior Dogs Supplemented with a Senolytic and NAD+ Precursor Combination."

47. Simon, Russell, Mondino, et al. "A Randomized, Controlled Clinical Trial Demonstrates Improved Owner-Assessed Cognitive Function in Senior Dogs."

48. Dockser Marcus, A., Janin, A. & Shifflett, S. "'Reverse Aging' Guru David Sinclair and His Trail of Failed Businesses." *Wall Street Journal* (2024). https://www.wsj.com/health/wellness/david-sinclair-reverse-aging-failed-business-8bc4a43d.

49. Science of Us. "Is Elysium Health's Basis the Fountain of Youth?"

50. Weintraub. "Rejuvenating the Chance of Motherhood?"

51. Brenner, C. A "Science-Based Review of the World's Best-Selling Book on Aging." *Archives of Gerontology and Geriatrics* 104, 104825 (2022).

52. Husten. "Resveratrol and Fraud."

53. Methuselah Foundation homepage. https://www.mfoundation.org/.

54. Methuselah Foundation homepage.

55. Methuselah Foundation homepage.

56. Methuselah Foundation homepage.

57. "Dr. de Grey Fitness for Duty Agreement." SENS Research Foundation. https://sens.org/dr-de-grey-fitness-for-duty-agreement/.

58. Akst, J. "Aubrey de Grey on Leave After Sexual Harassment Allegations." The Scientist (2021). https://www.the-scientist.com/aubrey-de-grey-on-leave-after-sexual-harassment-allegations-69081.

59. Deming, L. "Aubrey." Blog (2021). https://ldeming.posthaven.com/aubrey; Halioua, C. "Aubrey de Grey Is a Sexual Predator." https://www.celinehh.com/aubrey.

60. Akst. "Aubrey de Grey on Leave After Sexual Harassment Allegations."

61. Akst. "Aubrey de Grey on Leave After Sexual Harassment Allegations."

62. "Executive Summary of Investigative Findings." SENS Research Foundation (2021). https://web.archive.org/web/20220329000327/https://www.sens.org/wp-content/uploads/2021/09/SENS-Executive-Summary-For-Public-Release-FINAL-091021-00323152xC0E95.pdf.

63. SENS Research Foundation. "Executive Summary of Investigative Findings."

64. Deming. "Aubrey."

65. SENS Research Foundation. "Executive Summary of Investigative Findings."

66. SENS Research Foundation. "Executive Summary of Investigative Findings"; Molteni, M. "Aubrey de Grey Made Sexual Comments to Two Scientists, Investigator Finds." STAT News (2021). https://www.statnews.com/2021/09/11/aubrey-de-grey-inappropriate-sexual-comments-female-entrepreneurs-independent-investigation/.

67. SENS Research Foundation. "Executive Summary of Investigative Findings."

68. Akst. "Aubrey de Grey on Leave After Sexual Harassment Allegations"; "Statement from the SENS Research Foundation Board of Directors." SENS Research Foundation. http://web.archive.org/web/20210904150955/https://sens.org/statement-from-srf-board-of-directors/.

69. Deming. "Aubrey"; Halioua. "Aubrey de Grey Is a Sexual Predator"; Molteni. "Aubrey de Grey Made Sexual Comments to Two Scientists, Investigator Finds."

70. Deming. "Aubrey"; Halioua. "Aubrey de Grey Is a Sexual Predator"; SENS Research Foundation. "Dr. de Grey Fitness for Duty Agreement."

71. SENS Research Foundation. "Dr. de Grey Fitness for Duty Agreement."

72. "Zuzalu." VitaDAO. https://www.vitadao.com/projects/zuzalu.

73. Dodes, R. "Bryan Johnson's Antiaging Quest Has Made Headlines. But There's More to His Story." *Vanity Fair* (2023); Dodes, R. "Bryan Johnson, the Antiaging and Penis-Science Biohacker, Makes Claims About His Ex-Fiancée." *Vanity Fair* (2024). https://www.vanityfair.com/style/bryan-johnson-ex-fiancee-taryn-southern; Onose, G., Mureşanu, D. F., Ciurea, A. V., et al. "Neuroprotective and Consequent Neurorehabilitative Clinical Outcomes, in Patients Treated with the Pleiotropic Drug Cerebrolysin." *The Journal of Medicine and Life* **2**, 350 (2009).

74. Dodes. "Bryan Johnson's Antiaging Quest Has Made Headlines."

75. Bryan Johnson on X: "Men don't get abused, right?" https://x.com/bryan_johnson/status/1757800796476018918

76. Dodes. "Bryan Johnson's Antiaging Quest Has Made Headlines"; Bryan Johnson on X: "Ten facts for your reading pleasure." https://x.com/bryan_johnson/status/1757801140564136406.

77. Vance, A. "Bryan Johnson's Anti-Aging Blood Transfusion Involves Dad and Son." Bloomberg (2023). https://www.bloomberg.com/news/articles/2023-05-22/bryan-johnson-s-anti-aging-blood-transfusion-involves-dad-and-son?srnd=premium&leadSource=uverify%20wall&sref=MTy2GeXk.

78. Horrington, E. M., Pope, F., Lunsford, W. & McCay, C. M. "Age Changes in the Bones, Blood Pressure, and Diseases of Rats in Parabiosis." *Gerontology* 4, 21–31 (1960).

79. Scudellari, M. "Ageing Research: Blood to Blood." *Nature* (2015). http://web.archive.org/web/20200503204752/https://www.nature.com/news/ageing-research-blood-to-blood-1.16762#/b4.

80. Zou, S., Musavi, F., Notari IV, E. P. & Fang, C. T. "Changing Age Distribution of the Blood Donor Population in the United States." *Transfusion (Paris)* 48, 251–257 (2008); World Health Organization. *Global Status Report on Blood Safety and Availability 2021* (World Health Organization, Geneva, 2022), 1–184; Kracalik, I., Sapiano, M. R. P., Wild, R. C., et al. "Supplemental Findings of the 2021 National Blood Collection and Utilization Survey." *Transfusion (Paris)* 63 Suppl 4, S19–S42 (2023).

81. World Health Organization. *Global Status Report on Blood Safety and Availability 2021*.

82. World Health Organization. *Global Status Report on Blood Safety and Availability 2021*.

83. Jones, J. M., Sapiano, M. R. P., Mowla, S., et al. "Has the Trend of Declining Blood Transfusions in the United States Ended? Findings of the 2019 National Blood Collection and Utilization Survey." *Transfusion (Paris)* **61**, S1 (2021); Kracalik, Sapiano, Wild, et al. "Supplemental Findings of the 2021 National Blood Collection and Utilization Survey."

84. Bercovici, J. "Peter Thiel Is Very, Very Interested in Young People's Blood." Inc.com (2016). https://www.inc.com/jeff-bercovici/peter-thiel-young-blood.html.

85. Bercovici. "Peter Thiel Is Very, Very Interested in Young People's Blood."

86. Ehrenkranz, M. "Inside Ambrosia: Could Infusions of Millennial Blood Make You Young Again? Scientists Have Doubts." Mic (2017). https://www.mic.com/articles/179829/inside-ambrosia-could-infusions-of-millennial-blood-make-you-young-again-scientists-doubt-it#.2eTTBEZAB.

87. Ehrenkranz. "Inside Ambrosia."

88. Toner, R. W., Pizzi, L., Leas, B., et al. "Costs to Hospitals of Acquiring and Processing blood in the US: A Survey of Hospital-Based Blood Banks and Transfusion Services." *Applied Health Economics and Health Policy* 9, 29–37 (2011).

89. Bercovici. "Peter Thiel Is Very, Very Interested in Young People's Blood."

90. Liebelson, D. & Cook, J. "He Hawks Young Blood As A New Miracle Treatment. All That's Missing Is Proof." HuffPost (2018). https://www.huffingtonpost.co.uk/entry/ambrosia-young-blood-plasma-jesse-karmazin_n_5c1bbafce4b0407e9078373c.

91. Parental Consent Form Revision Eight. South Texas Blood and Tissue. http://southtexasblood.org; Cook, H. "Notorious 'Young Blood' Doctor Claims Dead Patient Faked His Own Death." HuffPost (2019). https://www.huffingtonpost.co.uk/entry/ambrosia-doctor-jesse-karmazin-false-claim-patient-faked-death_n_5d955bb6e4b0da7f66215584.

92. "Infusions of Young Blood Not Proven 'Safe or Effective,' US Government Warns." *The Guardian* (2019). https://www.theguardian.com/science/2019/feb/19/infusions-young-blood-not-proven-safe-effective-fda-warns.

93. Cook. "Notorious 'Young Blood' Doctor Claims Dead Patient Faked His Own Death."

94. Trotter, J. K. "Someone Is Trying to Discredit the Story of Peter Thiel's Interest in Young Blood." Gizmodo (2017). https://gizmodo.com/someone-is-trying-to-discredit-the-story-of-peter-thiel-1796135794.

95. "No, Peter Thiel Is Not Harvesting the Blood of the Young." TechCrunch (2017). https://techcrunch.com/2017/06/14/no-peter-thiel-is-not-harvesting-the-blood-of-the-young/.

96. Trotter. "Someone Is Trying to Discredit the Story of Peter Thiel's Interest in Young Blood."

97. Science of Us. "Is Elysium Health's Basis the Fountain of Youth?"

98. Cook. "Notorious 'Young Blood' Doctor Claims Dead Patient Faked His Own Death."

99. Montana Code Annotated, Title 50, Chapter 12, Part 1, "Right to Try Act" (2023).

100. Montana Code Annotated, Title 50, Chapter 12, Part 1, "Right to Try Act" (2023).

101. "About." a4li.org. https://a4li.org/about/.

Chapter 11

1. Weismann, A. *Essays Upon Heredity and Kindred Biological Problems*, vol. 1 (Clarendon Press, Oxford, 1891).

2. Weismann. *Essays Upon Heredity and Kindred Biological Problems.*

3. Weismann. *Essays Upon Heredity and Kindred Biological Problems.*

4. Carrel, A. "On the Permanent Life of Tissues Outside of the Organism." *Journal of Experimental Medicine* **15**, 516 (1912).

5. Swim, H. E. & Parker, R. F. "Culture Characteristics of Human Fibroblasts Propagated Serially." *American Journal of Hygiene* 66, 235–243 (1957); Hayflick, L. "The Limited In Vitro Lifetime of Human Diploid Cell Strains." *Experimental Cell Research* 37, 614–636 (1965).

6. Blackburn, E. & Epel, E. *The Telomere Effect: A Revolutionary Approach to Living Younger, Healthier, Longer* (Grand Central Publishing, New York, 2017).

7. Blackburn & Epel. *The Telomere Effect*; Harley, C. B., Futcher, A. B. & Greider, C. W. "Telomeres Shorten During Ageing of Human Fibroblasts." *Nature* 345, 458–460 (1990).

8. Aubert, G. & Lansdorp, P. M. "Telomeres and Aging." *Physiological Reviews* 88, 557–579 (2008); de Lange, T. "Telomeres and Senescence: Ending the Debate." *Science* 279, 334–335 (1998); Shay, J. W. & Wright, W. E. "Senescence and Immortalization: Role of Telomeres and Telomerase." *Carcinogenesis* 26, 867–874 (2005); Hahn, W. C. "Role of Telomeres and Telomerase in the Pathogenesis of Human Cancer." *Journal of Clinical Oncology* 21, 2034–2043 (2003).

9. Weismann. *Essays Upon Heredity and Kindred Biological Problems.*

10. Medawar, P. B. *The Uniqueness of the Individual* (Methuen & Co. Ltd., London, 1957).

11. Hayflick, L. "The Limited In Vitro Lifetime of Human Diploid Cell Strains." *Experimental Cell Research* 37, 614–636 (1965).

12. Aubert & Lansdorp. "Telomeres and Aging"; Shay & Wright. "Senescence and Immortalization"; Sharpless, N. E. & DePinho, R. A. "Telomeres, Stem Cells, Senescence, and Cancer." *Journal of Clinical Investigation* 113, 160–168 (2004).

13. Challis, G. B. & Stam, H. J. "The Spontaneous Regression of Cancer. A Review of Cases from 1900 to 1987." *Acta Oncology* 29, 545–550 (1990).

14. Challis & Stam. "The Spontaneous Regression of Cancer."

15. Challis & Stam. "The Spontaneous Regression of Cancer."

16. Challis & Stam. "The Spontaneous Regression of Cancer."

17. Moreno, E. "Is Cell Competition Relevant to Cancer?" *Nature Reviews Cancer* 8, 2, 141–147 (2008).

18. Moreno. "Is Cell Competition Relevant to Cancer?"

19. Minot, C. *The Problem of Age, Growth, and Death: A Study of Cytomorphosis* (G.P. Putnam & Sons., New York and London, 1908).

20. Naugler, C. T. "Population Genetics of Cancer Cell Clones: Possible Implications of Cancer Stem Cells." *Theoretical Biology and Medical Modelling* 7, 42 (2010).

21. Blasco, M. A., Lee, H. W., Hande, M. P., et al. "Telomere Shortening and Tumor Formation by Mouse Cells Lacking Telomerase RNA. *Cell* 91, 25–34 (1997).

22. Blasco, Lee, Hande, et al. "Telomere Shortening and Tumor Formation by Mouse Cells Lacking Telomerase RNA."

23. Rubin, H. "The Disparity Between Human Cell Senescence In Vitro and Lifelong Replication In Vivo." *Nature Biotechnology* 20, 675–681 (2002).

24. "Elizabeth Blackburn, PhD." Academy of Achievement (2000). https://achievement.org/achiever/elizabeth-blackburn/#interview.

25. Academy of Achievement. "Elizabeth Blackburn, PhD."

26. Blackburn & Epel. *The Telomere Effect*.

27. López-Otín, C., Blasco, M. A., Partridge, L., Serrano, M. & Kroemer, G. "The Hallmarks of Aging." *Cell* 153, 1194–1217 (2013).

28. Sharpless, N. E. & Sherr, C. J. "Forging a Signature of In Vivo Senescence." *Nature Reviews Cancer* 15, 397–408 (2015).

29. Sharpless & Sherr. Forging a Signature of In Vivo Senescence."

30. Jaijyan, D. K., Selariu, A., Cruz-Cosme, R., et al. "New Intranasal and Injectable Gene Therapy for Healthy Life Extension." *Proc Natl Acad Sci* 119, e2121499119 (2022).

31. Jaijyan, Selariu, Cruz-Cosme, et al. "New Intranasal and Injectable Gene Therapy for Healthy Life Extension."

32. Jaijyan, Selariu, Cruz-Cosme, et al. "New Intranasal and Injectable Gene Therapy for Healthy Life Extension."

33. Charles Brenner, PhD, on X: "note also this is a company that runs gene therapy trials abroad to skirt safety regulations and sells nonvalidated biological testing kits bad actors." https://x.com/CharlesMBrenner/status/1527857016567304192.

34. "New Intranasal and Injectable Gene Therapy for Healthy Life Extension." PubPeer. https://pubpeer.com/publications/9B76A71C2995A92B90D37AC1BE0693#3.

35. PubPeer. "New Intranasal and Injectable Gene Therapy for Healthy Life Extension."

36. Jaijyan, D. K., Selariu, A., Cruz-Cosme, R., et al. "Erratum: New Intranasal and Injectable Gene Therapy for Healthy Life Extension." *Proceedings of the National Academy of Sciences* 120, e2311483120 (2023).

37. Salisbury, I. "This Map Shows the Best-Paying Employer in Every State." Money (2018). https://money.com/best-paying-companies/.

38. Bodnar, A. G., Ouelette, M., Frolkis, M., et al. "Extension of Life-Span by Introduction of Telomerase into Normal Human Cells." *Science* 279, 349–352 (1998).

39. Moran, J. M. "Could Immortality Be Within Grasp? Scientists Say Yes." *Los Angeles Times* (1998). https://www.latimes.com/archives/la-xpm-1998-aug-24-he-16023-story.html.

40. Palca, J. "In Prelude to Nobel, Greider Talks Telomerase in Stockholm." NPR (2009). https://www.npr.org/sections/health-shots/2009/12/greider_nobel_lecture.html.

41. *Elizabeth H. Blackburn Papers*. Manuscript Division, Library of Congress, Washington, DC.

42. "Geron Reports Inducement Grants Under Nasdaq Listing Rule 5635(c)(4)." Geron Corporation (2024). https://ir.geron.com/investors/press-releases/press-release-details/2024/Geron-Reports-Inducement-Grants-Under-Nasdaq-Listing-Rule-5635c4-0bed34d97/default.aspx.

43. Stephen S. Hall. *Merchants of Immortality: Chasing the Dream of Human Life Extension* (Houghton Mifflin Company, Boston, 2003).

44. Hall. *Merchants of Immortality*.

45. Hall. *Merchants of Immortality*.

46. Hall. *Merchants of Immortality*.

Chapter 12

1. Huang, J. & Ganeri, J. "Is This Me? A Story About Personal Identity from the Mahāprajñāpāramitopadeśa/Dà zhìdù lùn." *British Journal for the History of Philosophy* 29, 739–762 (2021).

2. Huang & Ganeri. "Is This Me? A Story About Personal Identity from the Mahāprajñāpāramitopadeśa."

3. Huang & Ganeri. "Is This Me? A Story About Personal Identity from the Mahāprajñāpāramitopadeśa."

4. "Theseus: By Plutarch." The Internet Classics Archive. http://classics.mit.edu//Plutarch/theseus.html.

5. "De Die Natale." Censorinus. https://elfinspell.com/ClassicalTexts/Maude/Censorinus/DeDieNatale-Part1.html; Carpi, J. B. da. *Isagogae Breues et Exactissimae in Anatomia Humani Corporis* (Henricus Sybold, Strasbourg, 1530).

6. Vesalius, A. *De Humani Corporis Fabrica Libri Septem* (Joannes Oporinus, Basel, 1543); Carpi. *Isagogae Breues et Exactissimae in Anatomia Humani Corporis*.

7. Descartes, R. *Discourse on the Method of Rightly Conducting the Reason, and Seeking Truth in the Sciences*, vol. 4 (Open Court Publishing Company, London, 1899).

8. Descartes, R. *The Philosophical Writings of Descartes*, vol. 3 (Press Syndicate, Cambridge, 1997).

9. Descartes. *The Philosophical Writings of Descartes*.

10. Carpi. *Isagogae Breues et Exactissimae in Anatomia Humani Corporis*.

11. Vesalius. *De Humani Corporis Fabrica Libri Septem*.

12. Keating, J. *Animating Empire: Automata, the Holy Roman Empire, and the Early Modern World* (Pennsylvania State University Press, University Park, 2018).

13. Keating. *Animating Empire*.

14. Dawson, V. *Nature's Enigma: The Problem of the Polyp in the Letters of Bonnet, Trembley and Reaumur*, vol. 174 (American Philosophical Society, Philadelphia, 1987).

15. Williams, G. C. "Pleiotropy, Natural Selection and the Evolution of Senescence." *Evolution (NY)* 11, 398–411 (1957).

16. Orgel. "The Maintenance of the Accuracy of Protein Synthesis and Its Relevance to Ageing."

17. Klutke, G. A., Kiessler, P. C. & Wortman, M. A. "A Critical Look at the Bathtub Curve." *IEEE Trans Reliab* 52, 125–129 (2003); Rigdon, S. & Basu, A. *Statistical Methods for the Reliability of Repairable Systems* (John Wiley & Sons, New York, 2000).

18. Finch, C. *Between Zeus and the Salmon: The Biodemography of Longevity* (National Academies Press, Washington, DC, 1997); Gavrilov, L. A. & Gavrilova, N. S. "The Reliability-Engineering Approach to the Problem of Biological Aging." *Annals of the New York Academy of Sciences* 1019, 509–512 (2004); Gavrilov, L. A. & Gavrilova, N. S. "The Reliability Theory of Aging and Longevity." *Journal of Theoretical Biology* 213, 527–545 (2001).

19. Finch. *Between Zeus and the Salmon*.

20. Aubert, G. & Lansdorp, P. M. "Telomeres and Aging." *Physiological Reviews* 88, 557–579 (2008); Bokov, A., Chaudhuri, A. & Richardson, A. "The Role of Oxidative Damage and Stress in Aging." *Mechanisms of Ageing and Development* 125, 811–826 (2004); Brys, K., Vanfleteren, J. R. & Braeckman, B. P. "Testing the Rate-of-Living/Oxidative Damage Theory of Aging in the Nematode Model Caenorhabditis Elegans." *Experimental Gerontology* 42, 845–851 (2007). https://doi.org/10.1016/j.exger.2007.02.004; Campisi, J., Kim, S. H., Lim, C. S. & Rubio, M. "Cellular Senescence, Cancer and Aging: The Telomere Connection." *Experimental Gerontology* 36, 1619–37 (2001); Harman, D. "Aging: A Theory Based on Free Radical and Radiation Chemistry." *Journal of Gerontology* 11, 298–300 (1956); Gavrilov, L. A. & Gavrilova, N. S. "Reliability Theory of Aging and Longevity" in *Handbook of the Biology of Aging* 3–42 (2005). doi:10.1016/B978-012088387-5/50004-2; Sohal, R. S. & Weindruch, R. "Oxidative Stress, Caloric Restriction, and Aging." *Science* 273, 59 (1996).

21. Newman, S. J. "British Automobiles, Aging Theory, and the Death of Complex Machines." *ArXiv* (2025).

22. "ISO 3779:2009(en), Road Vehicles—Vehicle Identification Number (VIN)—Content and Structure." International Organization for Standardization. https://www.iso.org/obp/ui/#iso:std:iso:3779:ed-4:v1:en:sec:3.5; "ISO 3780:2009—Road Vehicles—World Manufacturer Identifier (WMI) Code." International Organization for Standardization. https://www.iso.org/standard/45844.html.

23. Finch. *Between Zeus and the Salmon.*

24. Gavrilov & Gavrilova. "The Reliability-Engineering Approach to the Problem of Biological Aging."

25. Finch. *Between Zeus and the Salmon.*

26. Newman. "British Automobiles, Aging Theory, and the Death of Complex Machines."

27. Newman. "British Automobiles, Aging Theory, and the Death of Complex Machines."

28. Finch. *Between Zeus and the Salmon.*

29. Gavrilov & Gavrilova. "The Reliability Theory of Aging and Longevity"; Gavrilov & Gavrilova. "Reliability Theory of Aging and Longevity."

30. Newman. "British Automobiles, Aging Theory, and the Death of Complex Machines."

31. "ISO 3779:2009(en), Road Vehicles—Vehicle Identification Number (VIN)—Content and Structure." International Organization for Standardization. https://www.iso.org/obp/ui/#iso:std:iso:3779:ed-4:v1:en:sec:3.5; "ISO 4030:1983(en), Road Vehicles—Vehicle Identification Number (VIN)—Location and Attachment." International Organization for Standardization. https://www.iso.org/obp/ui/#iso:std:iso:4030:ed-2:v1:en.

32. Newman. "British Automobiles, Aging Theory, and the Death of Complex Machines."

33. Newman. "British Automobiles, Aging Theory, and the Death of Complex Machines."

34. Newman. "British Automobiles, Aging Theory, and the Death of Complex Machines."

35. Newman, S. J., Schulte, K., Morellini, M. M., Rahal, C. & Leasure, D. R. "Offshoring Emissions Through Used Vehicle Exports." *Nature Climate Change* 14, 238–241 (2024).

36. "Used Vehicles Exported to Africa: A Study on the Quality of Used Export Vehicles." Netherlands Human Environment and Transport Inspectorate (2020). https://www.ilent.nl/binaries/ilt/documenten/rapporten/2020/10/26/rapport--used-vehicles-exported-to-africa/.

37. Newman, Schulte, Morellini, Rahal & Leasure. "Offshoring Emissions Through Used Vehicle Exports"; Netherlands Human Environment and Transport Inspectorate. "Used Vehicles Exported to Africa"; United Nations Environment Programme. *Used Vehicles and the Environment—A Global Overview of Used Light Duty Vehicles: Flow, Scale, and Regulation* (United Nations, Nairobi, 2020).

38. Lelieveld, J., Evans, J. S., Fnais, M., Giannadaki, D. & Pozzer, A. "The Contribution of Outdoor Air Pollution Sources to Premature Mortality on a Global Scale." *Nature* 525, 367–371 (2015); Chambliss, S. E., Silva, R., West, J. J., Zeinali, M. & Minjares, R. "Estimating Source-Attributable Health Impacts of Ambient Fine Particulate Matter Exposure: Global Premature Mortality from Surface Transportation Emissions in 2005." *Environmental Research Letters* 9, 104009 (2014).

39. Wentworth Thompson, D. *On Growth and Form* (Cambridge University Press, Cambridge, 1917); Rigdon & Basu. *Statistical Methods for the Reliability of Repairable Systems*; Gavrilov & Gavrilova. "Reliability Theory of Aging and Longevity."

40. United Nations Environment Programme. *Used Vehicles and the Environment*.

41. Storchmann, K. "On the Depreciation of Automobiles: An International Comparison." *Transportation (Amsterdam)* 31, 371–408 (2004).

42. Storchmann. "On the Depreciation of Automobiles"; Hamilton, B. & Macauley, M. "Competition and Car Longevity." *Resources for the Future Working Paper Series* 98, 1–28 (1998).

43. Newman, S. J. "Predicting the Evolution of Life History Using Inclusive Fitness." PhD thesis (Australian National University, 2015).

44. Hamilton & Macauley. "Competition and Car Longevity."

45. “The 2019 National Cost of Ownership Survey.” Federation of British Historic Vehicle Clubs (2019), 1–21. https://www.fbhvc.co.uk/uploads/files/2019%20FBHVC%20Cost%20of%20Ownership%20Summary.pdf.

Chapter 13

1. Belousov, B. P. “Периодически действующая реакция и её механизм [Periodically Acting Reaction and Its Mechanism].” *Сборник рефератов по радиационной медицине* [*Collection of Abstracts on Radiation Medicine*] 147, 145 (1959).

2. Minot, C. *The Problem of Age, Growth, and Death: A Study of Cytomorphosis* (G.P. Putnam & Sons, New York and London, 1908).

3. Mathieson, T., Franken, H., Kosinski, J., et al. “Systematic Analysis of Protein Turnover in Primary Cells.” *Nature Communications* 9, 1–10 (2018); Chen, W., Smeekens, J. M. & Wu, R. “Systematic Study of the Dynamics and Half-Lives of Newly Synthesized Proteins in Human Cells.” *Chemical Science* 7, 1393–1400 (2016); Geva-Zatorsky, N., Issaeva, I., Mayo, A., et al. “Using Bleach-Chase to Measure Protein Half-Lives in Living Cells.” *Nature Protocols* 7, 801–811 (2012).

4. Mathieson, Franken, Kosinski, et al. “Systematic Analysis of Protein Turnover in Primary Cells”; Chen, Smeekens & Wu. “Systematic Study of the Dynamics and Half-Lives of Newly Synthesized Proteins in Human Cells.”; Geva-Zatorsky, Issaeva, Mayo, et al. “Using Bleach-Chase to Measure Protein Half-Lives in Living Cells.”

5. Verzijl, N., DeGroot, J., Thorpe, S. R., et al. “Effect of Collagen Turnover on the Accumulation of Advanced Glycation End Products.” *Journal of Biological Chemistry* 275, 39027–39031 (2000).

6. Cohen, I. K. & Keiser, H. R. “Disruption of Healed Scars in Scurvy: The Result of a Disequilibrium in Collagen Metabolism.” *Plastic and Reconstructive Surgery* 57, 213–215 (1976).

7. Cohen & Keiser. “Disruption of Healed Scars in Scurvy.”

8. Rapoport, S. I. “In Vivo Fatty Acid Incorporation into Brain Phospholipids in Relation to Plasma Availability, Signal Transduction and Membrane Remodeling.” *Journal of Molecular Neuroscience* 16, 243–261 (2001).

9. Lajtha, A., Gibson, G. E. & Dienel, G. A. “Brain Energetics: Integration of Molecular and Cellular Processes” in *Handbook of Neurochemistry and Molecular Neurobiology: Brain Energetics. Integration of Molecular and Cellular Processes* (Springer, New York, 2007). doi:10.1007/978-0-387-30411-3.

10. Palacios-Marin, I., Serra, D., Jimenez-Chillarón, J., Herrero, L. & Todorčević, M. “Adipose Tissue Dynamics: Cellular and Lipid Turnover in Health and Disease.” *Nutrients* 15, 3968 (2023).

11. Arner, P., Bernard, S., Appelsved, L., et al. "Adipose Lipid Turnover and Long-Term Changes in Body Weight." *Nature Medicine* 25, 9, 1385–1389 (2019).

12. Parfitt, A., Travers, R., Rauch, F. & Glireaux, F. "Structural and Cellular Changes During Bone Growth in Healthy Children. *Bone* 27, 4, 487–494 (2000); Tsiridis, E. & Giannoudis, P. V. "Transcriptomics and Proteomics: Advancing the Understanding of Genetic Basis of Fracture Healing." *Injury* 37, S13–S19 (2006).

13. Yang, E., van Nimwegen, E., Zavolan, M., et al. "Decay Rates of Human mRNAs: Correlation with Functional Characteristics and Sequence Attributes." *Genome Research* 13, 1863 (2003).

14. Yang, van Nimwegen, Zavolan, et al. "Decay Rates of Human mRNAs."

15. Lindahl, T. "Instability and Decay of the Primary Structure of DNA." *Nature* 362, 6422, 709–715 (1993).

16. Allentoft, M. E., Collins, M., Harker, D., et al. "The Half-Life of DNA in Bone: Measuring Decay Kinetics in 158 Dated Fossils." *Proceedings of the Royal Society B: Biological Sciences* 279, 4724 (2012).

17. Lindahl, T. & Nyberg, B. "Rate of Depurination of Native Deoxyribonucleic Acid." *Biochemistry* 11, 3610–3618 (1972).

18. Haldane, J. B. S. *New Paths in Genetics* (George Allen & Unwin, London, 1941); Haldane, J. B. S. "The Rate of Spontaneous Mutation of a Human Gene." *Journal of Genetics* 31, 317–26 (1935).

19. Horvath, S. "DNA Methylation Age of Human Tissues and Cell Types." *Genome Biology* 14, R115 (2013).

20. Commerford, S. L., Carsten, A. L. & Cronkite, E. P. "Histone Turnover Within Nonproliferating Cells." *Proceedings of the National Academy of Sciences* 79, 1163–1165 (1982).

21. Gillis, J. A., Alsema, E. C. & Criswell, K. E. "Trunk Neural Crest Origin of Dermal Denticles in a Cartilaginous Fish." *Proceedings of the National Academy of Sciences* 114, 13200–13205 (2017).

22. Wilson, S. E. & Hong, J. W. "Bowman's Layer Structure and Function: Critical or Dispensable to Corneal Function? A Hypothesis." *Cornea* 19, 417–420 (2000); Wilson, S. E. "The Cornea: No Difference in the Wound Healing Response to Injury Related to Whether, or Not, There's a Bowman's Layer." *Biomolecules* 13 (2023).

23. Nowotny, K., Jung, T., Grune, T. & Höhn, A. "Accumulation of Modified Proteins and Aggregate Formation in Aging." *Experimental Gerontology* 57, 122–131 (2014).

24. López-Otín, C., Blasco, M. A., Partridge, L., Serrano, M. & Kroemer, G. "The Hallmarks of Aging." *Cell* 153, 1194–1217 (2013); Gavrilov & Gavrilova. "The Reliability

Theory of Aging and Longevity"; Bokov, A., Chaudhuri, A. & Richardson, A. "The Role of Oxidative Damage and Stress in Aging." *Mechanisms of Ageing and Development* 125, 811–826 (2004); Harman, D. "Aging: A Theory Based on Free Radical and Radiation Chemistry." *Journal of Gerontology* 11, 298–300 (1956); Gavrilov, L. A. & Gavrilova, N. S. "Reliability Theory of Aging and Longevity" in *Handbook of the Biology of Aging* (Academic Press, London, 2005), 3–42. doi:10.1016/B978-012088387-5/50004-2; Sohal, R. S. & Weindruch, R. "Oxidative Stress, Caloric Restriction, and Aging." *Science* 273, 59 (1996); Nowotny, Jung, Grune & Höhn. "Accumulation of Modified Proteins and Aggregate Formation in Aging"; López-Otín, C., Blasco, M. A., Partridge, L., Serrano, M. & Kroemer, G. "Hallmarks of Aging: An Expanding Universe." *Cell* 186, 243–278 (2023); Campisi, J. "Senescent Cells, Tumor Suppression, and Organismal Aging: Good Citizens, Bad Neighbors." *Cell* 120, 513–522 (2005); Harman, D. "Prolongation of Life: Role of Free Radical Reactions in Aging." *Journal of the American Geriatrics Society* 17, 721–735 (1969).

25. Höhfeld, J., Cyr, D. M. & Patterson, C. "From the Cradle to the Grave: Molecular Chaperones That May Choose Between Folding and Degradation." *EMBO Reports* 2, 885–890 (2001).

Chapter 14

1. Darwin, C. *On the Origin of Species by Means of Natural Selection, or the Preservation of Favoured Races in the Struggle for Life*, vol. 5 (John Murray, London, 1859), 63.

2. Fisher, R. A. *The Genetical Theory of Natural Selection* (Oxford University Press, Oxford, 1930).

3. Euler, L. "Recherches generales sur la mortalite et la multiplication du genre humaine." Manuscript (1760); Lotka, A. J. "Relation Between Birth Rates and Death Rates." *Science* 26, 21–22 (1907).

4. Hamilton, W. D. "The Moulding of Senescence by Natural Selection." *Journal of Theoretical Biology* 12, 12–45 (1966).

5. Hamilton. "The Moulding of Senescence by Natural Selection."

6. Hamilton. "The Moulding of Senescence by Natural Selection."

7. Williams, G. C. "Pleiotropy, Natural Selection and the Evolution of Senescence." *Evolution (NY)* 11, 398–411 (1957); Hamilton. "The Moulding of Senescence by Natural Selection."

8. Darwin. *On the Origin of Species by Means of Natural Selection*.

9. Johnstone, R. A. & Cant, M. A. "Evolution of Menopause." *Current Biology* 29, R112–R115 (2019).

10. Johnstone & Cant. "Evolution of Menopause."

11. Johnstone & Cant. "Evolution of Menopause"; Hawkes, K., O'Connell, J. F. & Blurton Jones, N. G. "Hadza Women's Time Allocation, Offspring Provisioning, and the Evolution of Long Postmenopausal Life Spans." *Current Anthropology* 38, 551–577 (1997); Ellis, S., Franks, D. W., Nielsen, M. L. K., Weiss, M. N. & Croft, D. P. "The Evolution of Menopause in Toothed Whales." *Nature* 627, 8004, 579–585 (2024); Hogendoorn, K. & Leys, R. "The Superseded Female's Dilemma: Ultimate and Proximate Factors That Influence Guarding Behaviour of the Carpenter Bee *Xylocopa Pubescens*." *Behavioral Ecology and Sociobiology* 33, 371–381 (1993); Uematsu, K., Kutsukake, M., Fukatsu, T., Shimada, M. & Shibao, H. "Altruistic Colony Defense by Menopausal Female Insects." *Current Biology* 20, 1182–1186 (2010); Foster, W. A. "Behavioural Ecology: The Menopausal Aphid Glue-Bomb." *Current Biology* 20, R559–R560 (2010); Wood, B. M., Negrey, J. D., Brown, J. L., et al. "Demographic and Hormonal Evidence for Menopause in Wild Chimpanzees." *Science* 282 (2023).

12. Ellis, Franks, Nielsen, Weiss & Croft. "The Evolution of Menopause in Toothed Whales"; Hogendoorn & Leys. "The Superseded Female's Dilemma"; Uematsu, Kutsukake, Fukatsu, Shimada & Shibao. "Altruistic Colony Defense by Menopausal Female Insects"; Foster. "Behavioural Ecology: The Menopausal Aphid Glue-Bomb."

13. Wood, Negrey, Brown, et al. "Demographic and Hormonal Evidence for Menopause in Wild Chimpanzees."

14. Hölldobler, B. & Wilson, E. O. *The Ants* (Harvard University Press, Cambridge, MA, 1990).

15. Oldroyd, B. P. & Osborne, K. E. "The Evolution of Worker Sterility in Honeybees: The Genetic Basis of Failure of Worker Policing." *Proceedings of the Royal Society B: Biological Sciences* 266, 1335 (1999).

16. Bennett, N. C., Faulkes, C. G. & Voigt, C. "Socially Induced Infertility in Naked and Damaraland Mole-Rats: A Tale of Two Mechanisms of Social Suppression." *Animals (Basel)* 12, 3039 (2022).

17. Schultheiss, P., Nooten, S. S., Wang, R., et al. "The Abundance, Biomass, and Distribution of Ants on Earth." *Proceedings of the National Academy of Sciences* 119, e2201550119 (2022).

18. Johnstone & Cant. "Evolution of Menopause"; Cohen, A. A. "Female Post-Reproductive Lifespan: A General Mammalian Trait." *Biological Reviews* 79, 733–750 (2004).

19. Skulachev, V. P. "Phenoptosis: Programmed Death of an Organism." *Biochemistry (Mosc)* 64, 1418–1426 (1999).

20. Hölldobler & Wilson. *The Ants*.

21. Foster. "Behavioural Ecology."

22. Hölldobler & Wilson. *The Ants*.

23. Strassmann, J. & Zhu, Y. "Altruism and Social Cheating in the Social Amoeba Dictyostelium Discoideum." *Nature* 408, 965–967 (2000).

24. Strassmann & Zhu. "Altruism and Social Cheating in the Social Amoeba Dictyostelium Discoideum."

25. Newman, S. J. "Predicting the Evolution of Life History Using Inclusive Fitness." PhD thesis (Australian National University, 2015); Newman, S. J. & Easteal, S. "A New Metric of Inclusive Fitness Predicts the Human Mortality Profile." *PLoS One* 10, e0117019 (2015).

26. Darwin. *On the Origin of Species by Means of Natural Selection*.

27. Haldane, J. "What Is Fitness" in *The Causes of Evolution* (Longmans, Green and Co., London, 1932), 111–143.

28. Maynard Smith, J. *The Theory of Evolution* (Penguin, Hardmondsworth, 1975). doi:10.2307/1931666.

29. Haldane, J. B. S. "Population Genetics." *New Biology* 18, 34–51 (1955).

30. Haldane. "Population Genetics."

31. Haldane. "Population Genetics."

32. Kimura, M. *The Neutral Theory of Molecular Evolution* (Cambridge University Press, Cambridge, 1983); Kimura, M. "Evolutionary Rate at the Molecular Level." *Nature* 217, 624–626 (1968); Kimura, M. "On the Change of Population Fitness by Natural Selection." *Heredity (Edinburgh)* 12, 145–167 (1958).

33. Wright, S. "Evolution in Mendelian Populations." *Genetics* 16, 97 (1931); Wright, S. "Volume II: The Theory of Gene Frequencies" in *Evolution and the Genetics of Populations*, vol. 2 (University of Chicago Press, Chicago, 1969), 511.

34. Wright. "Volume II: The Theory of Gene Frequencies."

35. Williams. "Pleiotropy, Natural Selection and the Evolution of Senescence."

36. Williams, P. & Day, T. "Antagonistic Pleiotropy, Mortality Source Interactions, and the Evolutionary Theory of Senescence." *Evolution (NY)* 57, 1478–1488 (2003).

37. Williams. "Pleiotropy, Natural Selection and the Evolution of Senescence."

38. Lee, R. "Rethinking the Evolutionary Theory of Aging: Transfers, Not Births, Shape Senescence in Social Species." *Proceedings of the National Academy of Sciences* 100, 9637–9642 (2003).

39. Orr, H. A. "Fitness and Its Role in Evolutionary Genetics." *Nature Reviews Genetics* 10, 531–539 (2009).

40. Grafen, A. A "First Formal Link Between the Price Equation and an Optimization Program." *Journal of Theoretical Biology* 217, 75–91 (2002); Grafen, A. "Formalizing Darwinism and Inclusive Fitness Theory." *Proceedings of the Royal Society of London. Series B, Biological Sciences* 364, 3135–3141 (2009); Grafen, A. "Optimization of Inclusive Fitness." *Journal of Theoretical Biology* 238, 541–563 (2006).

41. Hamilton, W. D. "The Genetical Evolution of Social Behaviour." *Journal Theoretical Biology* 7, 1–16 (1964).

42. Newman. "Predicting the Evolution of Life History Using Inclusive Fitness"; Newman & Easteal. "A New Metric of Inclusive Fitness Predicts the Human Mortality Profile."

43. Hamilton. "The Moulding of Senescence by Natural Selection."

44. Grafen. "Optimization of Inclusive Fitness"; Orlove, M. J. "A Model of Kin Selection Not Invoking Coefficients of Relationship." *Journal of Theoretical Biology* 49, 289–310 (1975).

45. Fisher. *The Genetical Theory of Natural Selection*; Grafen. "Optimization of Inclusive Fitness"; Orlove. "A Model of Kin Selection Not Invoking Coefficients of Relationship"; Price, G. R. "Fisher's 'Fundamental Theorem' Made Clear." *Annals of Human Genetics* 36, 129–140 (1972).

46. Grafen, A. "The Formal Darwinism Project in Outline: Response to Commentaries." *Biological Philosophy* 29, 281–292 (2014); Grafen, A. "The Formal Darwinism Project: A Mid-Term Report." *Journal of Evolutionary Biology* 20, 1243–1254 (2007).

47. Scott, T. W., Grafen, A. & West, S. A. "Multiple Social Encounters Can Eliminate Crozier's Paradox and Stabilise Genetic Kin Recognition." *Nature Communications* 13 (2022).

48. Hawkes, O'Connell & Blurton Jones. "Hadza Women's Time Allocation, Offspring Provisioning, and the Evolution of Long Postmenopausal Life Spans."

49. Sear, R. & Mace, R. "Who Keeps Children Alive? A Review of the Effects of Kin on Child Survival." *Evolution and Human Behavior* 29, 1–18 (2008); Sear, R., Mace, R. & McGregor, I. A. "Maternal Grandmothers Improve Nutritional Status and Survival of Children in Rural Gambia." *Proc Biol Sci* 267, 1641–1647 (2000); Sear, R., Steele, F., McGregor, I. & Mace, R. "The Effects of Kin on Child Mortality in Rural Gambia." *Demography* 39, 43–63 (2003).

50. Sear, Mace & McGregor. "Maternal Grandmothers Improve Nutritional Status and Survival of Children in Rural Gambia"; Sear, Steele, McGregor & Mace. "The Effects of Kin on Child Mortality in Rural Gambia."

51. Maynard-Smith, J. *Evolution and the Theory of Games* (Cambridge University Press, Cambridge, MA, 1982); Weibull, J. *Evolutionary Game Theory. Physica D Nonlinear Phenomena*, vol. 22 (MIT Press, Cambridge, MA, 1995); Nowak, M. A. "Evolving Cooperation." *Journal of Theoretical Biology* 299, 1–8 (2012). https://doi.org/10.1016/j.jtbi.2012.01.014; Nowak, M. A., Tarnita, C. E. & Wilson, E. O. "The Evolution of Eusociality." *Nature* 466, 1057–1062 (2010); Nowak, Tarnita & Wilson. "The Evolution of Eusociality"; Dawkins, R. *The Selfish Gene* (Oxford University Press, Oxford, 1976). doi:10.2307/2065678.

52. Newman. "Predicting the Evolution of Life History Using Inclusive Fitness"; Newman & Easteal. "A New Metric of Inclusive Fitness Predicts the Human Mortality Profile."

53. Fernández, J. & Toro, M. A. "A New Method to Estimate Relatedness from Molecular Markers." *Molecular Ecology* 15, 1657–1667 (2006). Queller, D. C. & Goodnight, K. F. "Estimating Relatedness Using Genetic Markers." *Evolution (NY)* 43, 258–275 (1989); Gusev, A., Lowe, J. K., Stoffel, M., et al. "Whole Population, Genome-Wide Mapping of Hidden Relatedness." *Genome Research* 19, 318–326 (2009).

54. Clapham, P. J. "Age at Attainment of Sexual Maturity in Humpback Whales, Megaptera Novaeangliae." *Canadian Journal of Zoology* 70, 1470–1472 (2011). https://doi.org/10.1139/z92-202.

55. Kaplanis, J., Gordon, A., Shor, T., et al. "Quantitative Analysis of Population-Scale Family Trees with Millions of Relatives." *Science* 360, 171–175 (2018).

56. Newman & Easteal. "A New Metric of Inclusive Fitness Predicts the Human Mortality Profile."

57. Newman & Easteal. "A New Metric of Inclusive Fitness Predicts the Human Mortality Profile."

58. Newman & Easteal. "A New Metric of Inclusive Fitness Predicts the Human Mortality Profile."

Chapter 15

1. Eiler, J. H., Masuda, M. M., Spencer, T. R., Driscoll, R. J. & Schreck, C. B. "Distribution, Stock Composition and Timing, and Tagging Response of Wild Chinook Salmon Returning to a Large, Free-Flowing River Basin." *Transactions of the American Fisheries Society* 143, 1476–1507 (2014).

2. Liao, J. C. "Neuromuscular Control of Trout Swimming in a Vortex Street: Implications for Energy Economy During the Kármán Gait." *Journal of Experimental Biology* 207, 3495–3506 (2004).

3. Eiler, Masuda, Spencer, Driscoll & Schreck. "Distribution, Stock Composition and Timing, and Tagging Response."

4. Robertson, O. H. "Prolongation of the Life Span of Kokanee Salmon (Oncorhynchus Nerka Kennerlyi) by Castration Before Beginning of Gonad Development." *Proceedings of the National Academy of Sciences* 47, 609–621 (1961).

5. Robertson. "Prolongation of the Life Span of Kokanee Salmon."

6. Robertson. "Prolongation of the Life Span of Kokanee Salmon."

7. Robertson. "Prolongation of the Life Span of Kokanee Salmon."

8. Finch, C. *Between Zeus and the Salmon: The Biodemography of Longevity* (National Academies Press, Washington, DC, 1997).

9. Lazenby-Cohen, K. A. "Communal Nesting in Antechinus-Stuartii (Marsupialia, Dasyuridae)." *Australian Journal of Zoology* 39, 273–283 (1991).

10. Fisher, D. O., Dickman, C. R., Jones, M. E. & Blomberg, S. P. "Sperm Competition Drives the Evolution of Suicidal Reproduction in Mammals." *Proceedings of the National Academy of Sciences* 110, 17910–17914 (2013).

11. Fisher, Dickman, Jones & Blomberg. "Sperm Competition Drives the Evolution of Suicidal Reproduction in Mammals."

12. Diamond, J. M. "Big-Bang Reproduction and Ageing in Male Marsupial Mice." *Nature* 298, 115–116 (1982).

13. Lazenby-Cohen. "Communal Nesting in Antechinus-Stuartii (Marsupialia, Dasyuridae)."

14. McAllan, B. M., Joss, J. M. P. & Firth, B. T. "Phase Delay of the Natural Photoperiod Alters Reproductive Timing in the Marsupial Antechinus Stuartii." *Journal of Zoology* 225, 633–646 (1991).

15. McAllan, Joss & Firth. "Phase Delay of the Natural Photoperiod Alters Reproductive Timing in the Marsupial Antechinus Stuartii."

16. Moore, G. H. "Aetiology of the Die-Off of Male Antechinus Stuartii." PhD thesis (Australian National University, 1974). doi:10.25911/5D6CF877894CD.

17. Zheng, A. & Pacala, S. W. "The Enigmatic Life History of the Bamboo Explained as a Strategy to Arrest Succession." *Ecological Monographs* 94, e1621 (2024).

18. Janzen, D. H. "Why Bamboos Wait So Long to Flower." *Annual Review of Ecology, Evolution, and Systematics* 7, 347–391 (1976).

19. Janzen. "Why Bamboos Wait So Long to Flower."

20. Edgeloe, J. M., Severn-Ellis, A. A., Bayer, P. E., et al. "Extensive Polyploid Clonality Was a Successful Strategy for Seagrass to Expand into a Newly Submerged Environment." *Proceedings of the Royal Society B* 289 (2022).

21. Zheng & Pacala. "The Enigmatic Life History of the Bamboo Explained as a Strategy to Arrest Succession."

22. Kuijt, A. F., Burton, C. & Lamb, C. T. "Effects of Bear Endozoochory on Germination and Dispersal of Huckleberry in the Canadian Rocky Mountains." *PLoS One* 19, e0311809 (2024).

23. Loukoianov, A., Yan, L., Blechl, A., Sanchez, A. & Dubcovsky, J. "Regulation of VRN-1 Vernalization Genes in Normal and Transgenic Polyploid Wheat." *Plant Physiology* 138, 2364 (2005); Trevaskis, B., Bagnall, D. J., Ellis, M. H., Peacock, W. J. & Dennis, E. S. "MADS Box Genes Control Vernalization-Induced Flowering in Cereals." *Proceedings of the National Academy of Sciences* 100, 13099–13104 (2003).

24. Watson, A., Ghosh, S., Williams, M. J., et al. "Speed Breeding Is a Powerful Tool to Accelerate Crop Research and Breeding." *Nature Plants* 4, 23–29 (2018).

25. Newman, S. J. & Furbank, R. T. "Explainable Machine Learning Models of Major Crop Traits from Satellite-Monitored Continent-Wide Field Trial Data." *Nature Plants* (2021); Newman, S. J. & Furbank, R. T. "A Multiple Species, Continent-Wide, Million-Phenotype Agronomic Plant Dataset." *Scientific Data* 8 (2021); Victor, B. Nibali, A., Newman, S. J., et al. "High-Throughput Plot-Level Quantitative Phenotyping Using Convolutional Neural Networks on Very High-Resolution Satellite Images." *Remote Sensing* 16, 282 (2024); Furbank, R. T., Silva-Perez, V., Evans, J. R., et al. "Wheat Physiology Predictor: Predicting Physiological Traits in Wheat from Hyperspectral Reflectance Measurements Using Deep Learning." *Plant Methods* 17, 1–15 (2021); Voss-Fels, K. P., Robinson, H., Mudge, S. R., et al. "VERNALIZATION1 Modulates Root System Architecture in Wheat and Barley." *Molecular Plant* **11** (2018). https://doi.org/10.1016/j.molp.2017.10.005.

26. Sallon, S., Solowey, E., Cohen, Y., et al. "Germination, Genetics, and Growth of an Ancient Date Seed." *Science* 13, 320 (2008).

27. Sallon, Solowey, Cohen, et al. "Germination, Genetics, and Growth of an Ancient Date Seed."

28. Huang, G., Qin, S., Zhang, S., et al. "Performance, Economics and Potential Impact of Perennial Rice PR23 Relative to Annual Rice Cultivars at Multiple Locations in Yunnan Province of China." *Sustainability (Switzerland)* 10 (2018); Zhang, Y., Huang, G., Zhang, S., et al. "An Innovated Crop Management Scheme for Perennial Rice Cropping System and Its Impacts on Sustainable Rice Production." *European Journal of Agronomy* 122, 126186 (2021).

29. Huang, Qin, Zhang, et al. "Performance, Economics and Potential Impact of Perennial Rice PR23"; Zhang, Huang, Zhang, et al. "An Innovated Crop Management Scheme."

30. Glover, J. D., Reganold, J. P., Bell, L. W., et al. "Increased Food and Ecosystem Security via Perennial Grains." *Science* 328, 1638–1639 (2010).

Index